For LaVonne Ann—
you fill my life with magic

Table of Contents

continued

continued

PERSONAL POWER COMPANIES

Prologue

Every once in while the world actually *does* change for the better. It is in such times that self-reliance triumphs over dependence, and initiative supplants "the way things have always been." For those riding the crest of the wave, it's a heady feeling. This is what's happening right now with renewable energy. Every day more and more people realize how sensible it is to harvest their own energy from the sun and wind, and in the process the giant power plants are belching just a little less smoke.

But it's not the time to get all starry-eyed—not yet, anyway. Like most revolutions, this one is driven as much by profit as ideals; as much by practical considerations as by earth-saving philosophies. Each is a necessary element: profit and practicality are the powerful motivators that fuel the fires of change, just as ideals and philosophies are the winds that keep them burning—hopefully—in the right direction.

This is not a save-the-planet book. I'm not much good at saving planets; I can't even figure out how to keep noxious, flatland weeds from invading my mountain meadow. No, this is a roll-up-your-shirtsleeves-and-install-a-productive-solar-and-wind-system book. My goal is simple: to show you what renewable energy is all about, so no one profits too much at *your* expense.

If you are reading these words, it's because you have developed more than a passing interest in renewable energy. Probably you are seriously considering installing some sort of system, and you'd like to know as much about it as possible before handing over hard-earned cash to someone who might've been a snake oil salesman in a bygone age.

The beauty of a book is that you can learn from someone who has learned the hard way, but has no desire to sell you thousands of dollars worth of components and equipment. That way, when it does come time to make the plunge, you'll know your money is being well spent.

In 2002 my wife, LaVonne, and I penned a rather large book titled *Logs, Wind and Sun* in which we detailed our two-year endeavor to build a log home and power it totally—electrically speaking—with homegrown power from the wind and sun. We drew heavily from our own inventions, successes, missteps and adventures, but, because most people's true-life experiences are seldom broad-based enough for those seeking a firm foundation of general knowledge, we thoroughly researched every aspect of the book to make it as useful as it was fun to read.

Since it was released, *Logs, Wind and Sun* has been well-received, both by log home enthusiasts and those involved in renewable energy. But, while most readers on the log home side of things thought the renewable energy aspect offered a fresh new direction, many on the other side of the fence found all the talk about log homes to be a trifle confining. Why not, they suggested, write a book that nixed the logs and emphasized renewable energy?

Why not, indeed?

Thus *Power With Nature* was born. It is a book written in two parts. Part one, *Dog of the Sun—Cat of the Wind*, is a primer; a primer disguised as a fable, to be precise. In this fable the human protagonist—me, as it turns out—finds himself caught up in the maelstrom of the renewable energy revolution. Though all he really dreams of are normal pets, the Perfect Woman, and a simple, functional solar pond-pump system, he finds himself in the vanguard of a nature-driven revolution.

That is the fable part. Along the way, as you may have guessed, he learns the rock-bottom basics of solar and wind energy systems. This is the primer part.

Why the fable?

Because for most of us, a good fable lodges itself in mental nooks and crannies that boring expositions of facts can't. Take Brer Rabbit, for instance. Of all the teachers I've had in my life, Brer Rabbit has stuck with me the longest. He was one hare who knew how to get out of a tight spot; a master of manipulation, a practitioner of the devilish powers every child craves to possess in all of their cabalistic intricacies.

I loved him so much because I grew up in a time when big kids had no compunction about beating up on little kids. It was the way of the world. A kid had to be tough or clever, or both, to survive. Brer Rabbit taught me that, when all else failed, a bluff was a pretty good alternative to getting pulverized. He also taught me that large thickets of thorny vegetation—Russian olive trees, in my case—were great places to slither away from big kids, after bombarding them with snowballs.

By the time Nature's ever-burgeoning dictates pulled my attention toward other things, my little Brer Rabbit book was so faded, wrinkled and dog-eared it was hardly readable.

Everyone has their own fables. Maybe for you it was *The Three Little Pigs*, or *The Little Boy Who Cried Wolf*. I personally learned the fundamentals of sound building practices from the former, and when to keep my mouth shut from the latter.

Where else can a kid learn these things?

No one remembers how they learned their multiplication tables, or how they learned to read and write; these are skills that were hammered into our hardly-willing little brains by rote—day after day, week after week, year after year. But every kid remembers *Little Red Riding Hood* in all its lurid splendor.

That's because fables are *fun*. They fire the imagination to the same degree that evil-eyed teachers douse it. They stick in our minds long after we forget the product of 9 times 13 (quick! what is it?), or if the vowel preceding a double consonant is long or short.

So, for every grownup among you who, like me, still appreciates the power of a good story, *Dog of the Sun—Cat of the Wind*, is a quaint but instructive yarn; one where animals act in unnatural ways, and Nature loves them all the more for it.

Then there's part two, *Beyond the Grid*, where we'll actually dig into the mountain of knowledge we touched on in the fable. Intimidating as the subject may seem, it really isn't. That's because—unlike Quantum physics, or English Lit—it all makes sense. Besides, it's a necessary element. As great a teacher as ol' Brer Rabbit might have been, he didn't know beans about readin', 'ritin', and 'rithmatic, and if Missus Horseface hadn't taught you and me all that stuff, you wouldn't even be reading this. So hang in there; you're about to learn a lot of wonderful things that will change your life for the better.

Beyond the Grid delves into every aspect of using available, off-the-shelf solar, wind and micro-hydro technology to power all or part of your home, whether you live in town hooked to the power grid, or so far out of town you can't even recall where the last power pole stands, lonely and unappreciated. All the components you will hear about in the fable will be covered in detail here. You'll learn how everything goes together; how each component works with all the others. Most importantly, you will be amazed at the natural logic of it all—you will actually feel like this renewable energy thing is a doable enterprise.

For those of you who are building new homes in remote places, I'll discuss design considerations, heating and cooling options, and the best ways to get water out of the ground and into your house. I consider many of the options available though, admittedly, I do a poor job of hiding my preferences.

So sit back and enjoy the book. And don't despair. I'm not an academic. Though I relied heavily on my formal education to produce *Power With Nature*, I'm really just a guy who graduated from the Academy of Fairy Tales and enrolled in the School of Hard Knocks.

— *Rex A. Ewing, 2003*

Preface to the 2nd Edition

As I dove into the revisions for the second edition of this book, I kept the old adage, *if it ain't broke, don't fix it*, firmly in mind. I had no intentions of changing anything that didn't need to be changed, or rewriting any parts of the text that are as valid today as they were three years ago when *Power With Nature* first came out.

But some things simply *had* to be revised. Many companies mentioned in these pages have since changed their names or their products, and some new players have entered the game. Grid-tied solar systems have grown in popularity and thus deserve a stronger mention. And, of course, I have heard from many readers of the first edition who said, "I really loved your book, but why didn't you discuss...(fill in the blank)...?" Good points, all. So I've added a lot of new things in this new edition. (Ever wonder how you could use sunlight and wind to make hydrogen to power your house? Keep reading.)

I've also beefed up the chapter on wind energy and included a thorough explanation of how solar cells really work. And, since we're charging a whole nickel more for this edition (or so I've been told), I have included profiles of a few of the solar-powered homes I've come across while interviewing people for my columns in *Log Homes Illustrated* and *Countryside* magazines.

And finally, for those of you wondering if I changed the fable at the front of book to reflect the growing popularity of grid-tied systems, the answer is: Nope; couldn't bring myself to do it. I'm an off-grid guy at heart and so the fable remains, untouched and untouchable. Besides, the critters wouldn't let me change it even if I wanted to.

— *Rex A. Ewing, 2006*

A Special Thank You to Ron and Gretchen Larson

The cover photo of a roof-mounted PV array was taken at the home Ron and Gretchen Larson during construction. Dr. Larson, Chair of the American Solar Energy Society, and his wife purchased the 800-square foot, 2002 Solar Decathlon winning home built by the University of Colorado and incorporated it into their new home near Golden, Colorado. To read about their PV system, please see page 156.
Photo: LaVonne Ewing

Dog of the Sun
Cat of the Wind

a primer disguised as a fable

ILLUSTRATIONS BY SARA TUTTLE

big mean yellow cat
black dog runs hot on its tail
old script, new ending

I thought I was waking up to a normal day. The sun was rising in its rightful place over the eastern foothills while a few broken clouds filtered warm rays into golden shafts of shimmering light. I could hear the usual swarm of hummingbirds outside my window, chattering and posturing for places at the feeder. And the impatient (and at that hour, cacophonous) clicking of the dogs' claws on the oak floor was a familiar signal that they were eager to go outside and do all the things that dogs naturally do after a night's incarceration. Nothing abnormal about that.

It was only after I tromped down the stairs—bleary-eyed and half-asleep—that I began to realize my perfectly normal day wasn't going to last. For starters, the clock on the microwave was blank, meaning that the power was out—again. Not that unusual, when I stopped to think about it. Just annoying.

The real problem was the cat; the big yellow one. The one with the attitude. Wild Willie From Outta Town. He was up to something; I could feel it.

Just as always, there were three dogs pacing the floor in front of the big glass door, looking alternately at me and whatever it was outside demanding their immediate attention. When Big Mick, the leader of the pack, saw me coming down the stairs, he pushed his way into position, since it was his birthright to always be the first out the door. Amiable Amy was second, and Newt, The Nubian Devil Puppy, was at the bottom of the pecking order. She hung back in an uncomfortable proximity to Willie the cat and waited for me to knee my way through the traffic jam of hair, teeth and claws to open the door. The other cat, Stinky the Spook, warily eyed the anxious foursome from the relative safety of the couch, where she no doubt planned to take a long, peaceful nap while everyone else—particularly her nemesis, Wild Willie—was out chasing rabbits, ghosts, velociraptors, wooly mammoths, or whatever else might wander into range.

Wild Willie, true to his name, is a born troublemaker. He has no respect for authority, or sense of fair play. If he were a fox, he'd be the bane of every farmer's hen house; if a 3rd grade student, he'd sneak tacks onto the teacher's chair and pelt the back of her head with spitwads. Being trapped in the body of a housecat, however, Willie has to sate his aggressive urges with frequent sneak attacks on poor Stinky, and occasional, unprovoked swipes at Newt. Mick and Amy, on the other hand, manage to stay above the fray, since they both learned long ago the best way to deal with Willie was to refuse to acknowledge he exists.

On this particular morning, Willie had a devilish look in his eyes. He was planning something, that much was clear. It was the same look he'd had when he got Newt to chase him into a culvert that wasn't quite as big as she was. It took two 5-pound cans of vegetable shortening, a prime cut of beef, and a long pole with a padded end to get

her out again. I shivered at the thought. Newt has a short memory, and I was out of shortening.

But, what the heck? I needed a shower and a cup of coffee and the dogs needed out. Whatever that miscreant of a cat had planned, there wasn't much I could do to stop it. I opened the door and let the river of semi-domesticated predators pour out into the unsuspecting day.

Fortunately, the power outage was short lived, so after a few minutes I was able to take a shower and give myself an infusion of Java before setting about to do my morning chores: feeding and watering the horses and calves, and mucking the horse stalls. It's not a lot of fun, but it sure beats a real job.

Keeping a suspicious eye glued to my yellow cat, I could see that my first feelings about Wild Willie were on the money. The ornery cat had a real attitude on today. He acted like he had a tick buried in his backside and was convinced that Newt was the cause of it. He ambushed her at the horses' water trough, and again in the feed room. I could see the poor dog's patience was running thin.

Finally, just as I was making my way back to the house for some breakfast, Willie launched himself off the haystack and sunk his fish-hook claws into Newt's back. She took off like a rank bronc with Willie—ears laid back and a sardonic sneer on his face—determined to stay on for the full 8 seconds. It didn't last that long. Newt took a roll after a few strides, and Willie leapt off just before becoming two-dimensional. It was a footrace from then on.

Newt lit after Willie like a dog possessed. She chased him under the deck, around the house and clear down to the barn. I remember thinking *for Heaven's sake, just stay away from the culvert!*

Back up to the house, and around the fish pond. Then Willie ran up the tree, 50 yards downhill from the brand-spanking-new solar panel I'd bought to run the pump that aerates the pond water. I could swear I heard the cat chuckling as Newt growled and paced around the base of the old Ponderosa. Fine, I thought, it'll do the miserable cat some good to spend a day in the tree. But Willie had other plans. Just as I was turning to go in the house, he jumped out of the tree and ran right for the solar panel, Newt's hot, steamy breath ruffling the fur on his tail. I didn't want to see what was about to happen, but I could hardly turn away. I yelled at Newt to stop, but she goes stone deaf when her blood's up and this is one time I couldn't blame her. I didn't want to have to call the vet to extract my dog from a solar panel, but at that point I couldn't see any way around it.

Then, just as I predicted, Willie ran under the solar panel, while Newt ran right into it. I cringed, waiting for whatever god-awful noise is supposed to occur when a dog running full-bore meets a fixed panel of tempered glass, but to my utter amazement there was not a sound to be heard.

Nor was there a dog to be seen. Newt had vanished.

black dog in shadows
yellow dog unloads his mind
denial ensues

J ust the week before, I had installed that simple little solar system (not to be confused with the big one with planets and moons and all— mine was a lot smaller). It wasn't much, just a modest 30-watt solar panel, and a 12-volt pond pump, both purchased from one of those environmentally oriented mail-order houses. The idea was simple; whenever the sun shines, the pump pulls a column of water from the bottom of the pond and dribbles it over a pile of rocks on the shore where it drains back into the pond.

When the sun was low in the sky, the flow was ponderously feeble, but it got to be respectable around midday. Upon seeing that it actually worked, I just had to ask the question: "How does a blue, glass-covered rectangle of silicon make water flow over a waterfall?" The answer was simple: "Who cares? I'm a rancher, not a geek." It was true. I am a rancher. And I didn't care.

At least not then.

But now, considering that my solar panel had just swallowed my dog whole, I suddenly had a keen interest in solar power.

I stared in disbelief at the seemingly-innocuous "thing" that had just eaten Newt. I was already starting to miss her. Even Willie seemed a little disheartened by the whole thing. He sat on his haunches and tapped the hard surface of the solar panel with his paw, as if he expected Newt to somehow pop out of it and get back into the game. I shared his sentiment.

But like 'em or not, facts are facts and the dog was gone. The ominous question was: gone where? It wasn't like there was a bone-rending crash, or a crack in the solar panel, or any evidence at all that she had ever been near the thing. Could I have imagined it all? Maybe, I surmised, late nights rocking with the Rolling Stones at full volume had at last begun to alter my sense of reality. Newt was probably on the other side of house, trying to flush a rabbit out of a bush.

As much as I wanted this to be true, the thought of giving up the Stones for Barry Manilow—or worse yet, the Carpenters—gave me an awful sinking feeling in the pit of my stomach.

After a thorough search, however, it became clear that Newt was gone. She wasn't looking in the bushes and rocks for rabbits or chipmunks, and she wasn't busy trying to uproot voles. She was just plain gone. At least (I tried to console myself) I didn't have to rewire my brain to appreciate mushy music.

Then, just as the sun set and I was preparing to feed my two remaining dogs, there came a scratching at the door. It was Newt. I quickly let her in and examined her from head to claw. Not a scratch or bump; no singed hair, loose or missing teeth, or even any sore spots. But for all that, she still looked like she'd chased a deer to the Continental Divide and back. Her eyes were glazed over and her tongue dropped nearly to the floor. Maybe that was it! I mused. Maybe she'd taken after a deer and disappeared into the woods where I couldn't find her. About then *Mandy* started playing in the back of my head, so I desperately looked for some evidence—any evidence—that Newt hadn't spent the day chasing deer. Wouldn't the pads on her feet be worn or torn? I pulled up a foot and studied it. It was as clean and smooth as if she'd been napping on the floor for the past 12 hours. The other three feet were the same way. *Mandy* thankfully faded into *Angie* as I watched Newt wolf down her supper.

Time has a way of mellowing weirdness, so by the time I hit the sheets I had pretty well convinced myself that I hadn't really seen Newt swallowed by a hungry solar panel. Probably just some trick of the sunlight reflecting from the solar panel and nearby rocks; a sort of natural hologram. With that comforting—and completely erroneous—thought in mind, I drifted off to sleep.

Some time later (I can't say just when) I was awakened by a soft, whispering voice that seemed to be coming from the floor beside my bed. I instinctively reached for the light switch, but the voice said, "Don't. It's better that we do this in the dark. It'll be easier for you to deny we ever had this conversation."

My hand continued toward the light switch, anyway, but then I remembered what a comfort deniability had been to me all day and

I pulled it back. "Who are you?" I asked.

"What do you mean 'who?' It's me. Mick. Your main dog." He sounded offended.

"Of course," I answered lamely. "How silly of me."

"Look. This isn't easy for me, either," he admitted, "but it's the way it's got to be."

I strained to see my talking dog in the dark, but the night was pristinely black. "Okay," I said. "What's going on?"

"It's Newt. She wanted me to tell you something."

"Why not just tell me, herself?" I wondered aloud.

"Don't be ridiculous! Newt can't speak English; she barely understands it."

True enough. "Okay. What's on her mind?" I decided to play along, since by now I was convinced I was dreaming.

"She says the pond pump wants a battery."

"She said WHAT?"

Mick ignored my outburst and continued, "She says the pump has to work too hard when the sun's low, so it wants a battery to store the extra power the solar panel gives off when the sun is high. To even things out, so to speak."

"I didn't know Newt knew so much about electricity," I answered, trying to stifle a laugh.

Mick replied, "Actually, she's had a very keen interest in the subject ever since you tried to electrocute her with that awful electric fence."

"I never....!"

"I know, I know," he interrupted. "Don't worry. She's almost forgiven you."

Almost forgiven me, I silently mouthed. I shuddered to think what would happen if Willie ever got tangled up in the electric fence I'd strung around the calf pasture. Vengeance, real or imagined, was the cat's purpose for living.

"I've got that old tractor battery in the shed," I heard myself say.

"She knew you'd say that. She said to tell you not to be a cheapskate and go buy a good deep-cycle battery. It will keep the pump happy."

I had to ask: "How does she know so much about what the pump wants?"

"After running through it a few million times she got to know it pretty well," Mick answered, matter-of-factly.

"You mean.....?"

"Yeah. She spent the whole day riding the circuit from the panel to the pump, to the panel, and back to the pump. Wore the poor little thing out. Luckily, she was able to slip away, once the sun went down and the electrical potential of the system dropped to zero." Silence. Then, "So buy a battery, why don't you?"

"The electrical potential of the system?" I echoed. "Mick, where in the world did you learn...?"

"The Internet, after you go to bed. Why do you think I sleep all day?"

"I just thought you were lazy," I answered truthfully.

"Hmmmph. Figures."

"Sorry," I mumbled.

"Sorry enough to buy me some liver while you're in town?"

"Liver?" I repeated incredulously.

"And I want it grilled, not fried."

"Yeah. Right."

"With turkey gravy."

"Don't press your luck, mutt." I waited for a prickly retort, but there was only the fading sound of his breathing. "Hey, Mick...?" I fell silent when I heard his familiar, scratchy footsteps on the stair. 🌿

 a new battery
a highly-charged winding path
black dog's déjà vu

It was all a dream, I told myself the next morning as I drove to Pickett's Farm & Ranch Supply on the other side of the mountain. Just one long deeply weird dream. So why was I driving 20 miles to buy a battery? It wasn't like my pets had cornered me that morning, demanding with low growls and barred teeth I buy a battery for the stupid overworked pond pump. They all acted just like yesterday had never happened. Willie, though maybe a little subdued, still had that caustic gleam in his yellow-green eyes, and Newt bounded out the door as if she expected a dozen rabbits to be lined up on the deck waiting for the chase.

Even Mick wagged his tail and looked up at me with his simple, adoring gaze when I walked down the stair. And when I ruffled his fur and said, "So what's all this about electrical potential?" he cocked his head to one side, as if to say, "Huh?"

No, I lied to myself as I negotiated a steep switchback, I was buying a battery simply because it would increase the life of the pond pump (though, admittedly, I had no idea why).

I reached for a Stones tape, thought better of it, and stuck in James Taylor instead. It was soothing. Mud Slide Slim lives in a really simple world. Probably doesn't have dogs and cats.

Although the morning was cool, I rolled down both windows in the pickup and let the chilly breeze wash across my face and fill my nostrils with the refreshing scent of pine. Things seemed agreeably normal just then, and normal was exactly what I wanted. No more disappearing dogs. Certainly no more talking dogs.

It had to be the Stones, I concluded. That, and green chili burritos. And Mexican beer. It was time for a few changes in my lifestyle. From now on it would be mellow folk music, whole grain bread, organic salads and herbal tea.

Yeah, right.

Who was I trying to kid? I wondered, as a wave of nausea washed over me. That kind of stuff is poison to a guy like me.

So, rather than give in to a slow and excruciating death-by-granola, I resolved then and there to figure this whole thing out, even if it meant I had to learn something about electricity. It couldn't be all that bad; certainly a far simpler subject than dog and cat psychology. After all, it's just so much wiring; the wiring of a pond pump had to be less complicated than the wiring of vengeful cat's twisted brain.

Mick had said—*hold on a minute!* In my *dream*, Mick had said—that Newt was able to get out of the system once the sun went down and the electrical potential of the system dropped to zero. I wasn't exactly certain what electrical potential was, but I was pretty sure the battery I was about to buy would be full of it. So what would happen if Newt got stuck in the new system? Poor dog would never get out.

Fearing for Newt's life, I almost turned around. Then I realized just how silly I was being. I ejected James Taylor, stuck in a Rolling Stones tape, and drove to town.

I didn't buy liver.

Home again, I lugged the battery over to the pond and went to work. Mick expectantly sniffed the sack filled with all the other things I'd bought for the job at hand. He huffed derisively, cast me a cold stare and stomped off when he realized it was nothing but electrical stuff.

Had to be my imagination.

The guy at Pickett's Farm & Ranch Supply—Eli J. Pickett, "hisself," as it turned out—didn't know beans about solar panels, but he had batteries down to a science and seemed to know a fair bit about wiring things. Like tractors and hay swathers and heaters for stock tanks. "Can't just hook it all together and hope for the best," he assured me. "You gotta have switches and fuses. And a ground rod; gotta have one o' them."

"Why is that?" I had to ask.

"Don't know much about this sorta thing, do ya, Sonny?" he said, narrowing his eyes. "Didn't learn nuthin' in school, did'ya?"

"I was a History major," I answered, a bit defensively.

"Uh-huh." He said this as if I'd just told him I was a cross-dresser studying ballet. "Whatever. Here's how it all works...."

He explained that the copper ground rod had to be pounded into the ground and hooked to the battery, so "all them little 'lectrons" had a place to go, if there was a short in the system, or a lightning strike.

"So why do I need the fuses, then?" I asked.

"Them 'lectrons is funny," he explained. "Don't always know where they're gonna go. Anyway, the fuses are more for when they run backwards."

"Backwards?"

"Yep. How much did your little pump cost you?" he asked.

"About twenty bucks," I told him.

"And that solar panel thingy?"

"About two-hundred," I said.

"So what's to happen when water leaks into your cheap little pump's wiring and shorts it out?" I opened my mouth to speak, but he answered for me. "It's gonna send a surge of 'lectrons—supplied by this here battery..." he patted the battery as though it were his personal attack-trained rottweiler, "...back to that expensive solar gizmo, that's what. And all them teeny little sissy circuits inside it are gonna get fried." He made a sizzling, snapping noise between his tongue and his teeth, to make his point. "Same thing with the battery. If it'd short out....well, you think it through, yourself. Thing is, if sumthin' goes haywire the fuses keep it from spreading the misery, if ya get my drift."

I did. But what about the switches?

He pulled off his greasy John Deere ball cap, scratched his shiny, liver-spotted head, and said, "Boy, you're lucky you ran into me." I personally had my doubts, but I let him continue. "You gotta have a switch between the battery and the pump so you can turn the pump off at night and not run all the juice outta the battery. And you need a switch between the solar panel and the battery to keep from pumping too much juice into the battery when you ain't runnin' your pump all dang day. And just so you know how much juice your batteries got in 'em, you need a voltmeter."

"This is getting complicated," I complained. "Isn't there some way all this can be done automatically, so I don't have to be flipping switches and testing voltages all the time?"

"Well," he said, looking off into the distance. "Mebbe you could figure out a way to wire in a pickup's voltage regulator to keep the battery from takin' on too much charge, but ain't no way to stop that little pump from runnin' it all out again, 'least without a switch. That's why tractors and trucks got ignition switches."

"Come on," I objected. "There has to be something."

He looked at me as though I'd just insulted his intelligence. "Sure thing, Boy," he said with a mean little smirk. "Just about the time the sow sprouts wings and takes to the sky."

I bought the ground rod, fuses and switches, and even the voltmeter Eli assured me I needed ("Wouldn't drive your pickup very far without a gas gauge, now would'ya?"), but passed on the voltage regulator. It just didn't seem right to wire a car part into my solar pond pump setup. Not that I'd have known how to do it, anyway.

For a few days after that, everything worked perfectly. I turned on the pond pump every morning after the sun came up and the battery was taking on a good charge, then turned it off at night after the battery voltage began to drop. I was really kind of enjoying monitoring the little system with the voltmeter Eli had sold me. (Actually, it was a multimeter, since it also measured amps, ohms and circuit continuity, but at that point, voltage was the limit of my expertise with the thing.) I'd had no more late-night conversations with Mick, though he had taken to looking in his supper dish each night with an audible sigh of disgust. Even my conniving cat, Wild Willie, was behaving himself. For the most part. Oh, he still terrorized Stinky—poor little thing—whenever the notion struck, and he scrawled his name on Newt's nose from time to time, but there didn't appear to be much forethought to his acts of aggression. Which was good. There's nothing worse than a calculating feline.

Then one day the harmony I hoped would last forever became dangerously unbalanced. The day had begun with heavy clouds in the east, so I didn't bother to turn on the pump, deciding it would be better to wait until the clouds burned off and the sun began charging the battery. But by the time the sky cleared—about 9:00 o'clock, or thereabouts—I was busy sorting calves and forgot to turn the thing on. By late afternoon when I finally got around to checking it, the battery was sizzling and boiling. I didn't need a fancy meter to tell me it was overcharged. I turned off the solar panel and turned on the pump, making a mental note to turn it off before bedtime.

Unfortunately, I forgot to attach my mental note to a mental clock and let the pump run all night. Well, not quite all night, since Mick woke me up at 4:00 a.m. to let me know we had a problem. "Better wake up," he said, matter-of-factly.

I'd been dreaming about the Perfect Woman. She was just beginning to turn her head so I could finally get a good look at her face (so I'd know her when I actually met her in the flesh) when I was so thoughtlessly interrupted. "Don't tell me the pump's unhappy again," I managed to hiss through gritted teeth.

He answered, "Not at all. The pump is having a great time, at the battery's expense."

Then it hit me: I'd left the pump on all night. "Oh, my God! I need to turn the pump off!" I exclaimed. I started to roll out of bed, but Mick quickly put a paw on my chest and said, firmly, "Just keep your shorts on. Everything's under control; the battery's fine. Besides, we need to talk Newt out of the system, first."

Talk Newt out of the system? This had to be another bad dream; best just to play along for awhile. "What's Newt doing back in the system?" I asked. "She was sleeping beside the couch when I went to bed. Besides, the solar panel is turned off. She couldn't get into the system if she wanted to."

"For starters," Mick explained, "you left the pet door open, so we all went outside after you went to bed. Then that meddlesome yellow cat of yours flipped the switch to the solar panel. With the pump turning and the battery running low, it sucked Newt in like a bird into a jet engine."

"Why does this always happen to Newt?" I lamented.

"Hard to say," Mick answered, reflectively. "Could be Karma, or maybe just a love of cheap thrills. Personally, I think it's an extra-dimensional thing. But don't worry; Amy's out there with her. We'll get her back."

"How?" I asked. "We'll have to run the battery dry to do it. You know—that electrical potential business you were talking about."

"Only if it were a perfectly closed system," Mick assured me. "But

it's hardly that. There're power leaks everywhere; in the pump, the battery, the wires and ground rod, and even back through the solar panel. She'll be out again in no time."

Even though it was just a dream (I kept telling myself) I breathed a sigh of relief. I asked, "So why did you wake me up?"

"To tell you that you need to add a couple more components to the system to keep this from happening again."

"Such as....?"

"A charge controller, for starters. And a load controller."

"Is a charge controller like a voltage regulator?" I asked, recalling my conversation with Eli J. Pickett.

Mick replied, "Yeah, I guess you could say that. In the same way a computer is like an abacus."

"And a load controller.....?" I was beginning to get the gist of what he was talking about. It must be something to shut the pump off when the battery voltage dropped too low. "Wait a minute," I objected. "Eli said there wasn't any such thing!"

Mick was losing his patience. "Your friend and mentor, Eli," he retorted, "learned everything he'll ever know about electricity chasing down shorts on a '52 Massey Ferguson loader tractor."

Probably true.

"Anyway," he continued, "just add those two things to the system and we'll never have to have one of these conversations again."

"Promise?" I asked, hopefully. Late night conversations with a dog that knew more about electricity than I did were beginning to get a little unnerving. Not to mention dog-eating solar panels and switch-flipping cats. I was ready for things to get back to normal.

"At least until you decide to start messing with the system again," he warned. "Like when you get around to upgrading to AC."

"AC?"

"You know: alternating current—the kind all of civilization uses, where the current changes polarity about 60 or so times a second."

"Why would I do that?" I wondered.

"Never mind. You'll find out, soon enough."

"Why do I have the feeling you know something I don't?"

"Because dogs always know more about the future than people," he smugly assured me. "You're all too busy thinking with those big brains of yours to pay attention to what's happening around you."

I'd heard that before, but never from a dog. Finding myself short on witty replies, I waited for him to say more, but he didn't. I thought the conversation was over. I had just laid back in bed and closed my eyes, when, at last, Mick said, "By the way—didn't you forget something the last time you went to town?"

"Liver?" I asked, a bit sheepishly.

"It would do an old dog good, you know."

With that, he was gone. 🌿

my bittersweet dreams
harbingers, I quickly find
though never quite grasped

As I lay there wide awake, I considered getting up to observe the slow, laborious process of extracting Newt from the pond pump PV system. At the very least it would make for an interesting chapter in my memoirs. But then I thought better of it. So far, in the comforting refuge of denial, I'd managed to keep my sanity; were I to go outside and find Amy and Mick conjuring Newt out of the system, one wisp of charged dog particle at a time, while Willie looked on with impish amusement...well, it just might push me over the edge.

Instead, I rolled over and tried to will myself to sleep, a process that's usually about as successful as telling yourself a root canal is a painless procedure. I tried counting sheep, but gave up when the lambs bounding over the short little fence morphed into disturbingly familiar black dogs jumping into voracious solar panels. Finally, I concentrated on the dream I'd been enjoying before being rudely awakened by my know-it-all yellow dog. Amazingly, it worked. I picked up the dream right where it left off.

There she was, the Perfect Woman, right before my eyes. A little hazy at first, but quickly the dreamy veil lifted and I saw her with crystal clarity: tall and slender with long, wavy, light-brown hair falling delicately over her shoulders. If only she would turn around so I could see her face...and then, as if granting an ardent wish, she did.

I was immediately captivated by her mischievous, gray-green eyes. They seemed to magnify light and brighten the rest of her lightly-tanned face. She smiled demurely when she saw me staring at her, and I knew with painful certainty I was in love.

But where was I? I had to know where to look for her. With great effort, I pried my eyes from her enchanting face and tried to focus beyond. At first the background looked snowy, like a TV screen with no reception, but I persisted and soon saw long, black loops, pinched in the middle by open-ended cardboard scabbards, hanging from a wall; they looked like...fan belts? The Perfect Woman was in an auto parts store?

I shook my dream head to clear up my dream vision, but in the process I bumped my non-dream head on my non-dream headboard and woke myself up. Though it seemed I'd only been asleep for a minute or two, the sun was high above the eastern hills and I could hear the horses neighing for their breakfast. So be it. It was a far-fetched dream, anyway. A woman that beautiful wouldn't be caught dead in a parts store. What would be the point? There wasn't a guy in the world who wouldn't rebuild her whole drive train for one of her smiles, and pay for the parts, to boot.

I rolled out of bed, slipped into yesterday's dirty clothes, and plodded down the stairs. Stinky was sleeping blissfully on the couch; not exactly a news flash. All the other animals were already outside, so I pulled on my boots and stepped out the door to witness the mayhem firsthand.

Mick, Amy and Willie were all resting on the ground by the pond, their attention riveted by Newt, who was running around in circles like a dog who'd just survived a near-death experience in an ocean of amphetamines. Her long, black hair stood straight out from her body and her tongue flopped from side to side in a mouth that was twisted into a hyena-like leer. She ran to one side of the pond, hit the brakes, spun around three times in a quick—but graceful—four-legged pirouette, then zipped back around to the other side. Puppy "crazies" taken to a new order of magnitude.

The pond pump was running—kind of slow, I thought—and the switch to the solar panel was in the 'on' position. I cast a suspicious glance at Willie, who looked back at me with a smug stare, as if to say, "You'll never prove a thing." He was right.

I thought about testing the voltage in the battery—right before I considered testing the voltage in the dog—but decided it didn't matter. The sun was high, the sky clear, and the dog was alive and (a little too) well. Instead, I fed the horses and pumped some water for the calves. At least I had normal livestock.

A trip to Big City later that day yielded the two crowning jewels in my pond pump PV system: a charge controller and a load controller. Actually, I soon discovered, the one is the same component as the other. I just had to program each one a little differently.

The guy at the Super Solar Megastore was a clean-cut young techie-type named Gordon. He was the antithesis of Eli J. Pickett. Patient and ever mindful of my technical limitations, he explained in detail how everything worked and why it was needed. In fact, his delivery was so smooth I concluded he must spend a lot of time talking to ignoramuses like me.

"Actually," he explained, "a charge controller is really just a very sophisticated on/off switch. It senses the battery's state of charge and adjusts the amperage going into it as needed. Its purpose is to keep the battery charged without overcharging it. When the battery is charged, it allows just enough current through it to run the load—your pump. Same thing when you use it as a load controller. When the load draws more current than the battery can safely give, it shuts off the load until the battery reaches a safe state of charge, again."

He paused and scratched his head, then regarded me with a quizzical expression. "I've got to tell you: this is a lot of trouble—not to mention expense—for a simple little pond pump system. Why not just get a couple of cheap toggle switches and inline fuses, and then monitor the system with a voltmeter?"

Without thinking, I said, "I tried that. My dog didn't like it."

Unable to stifle a chuckle, Gordon raised an eyebrow and said, "Well, of course, if your dog didn't like it..."

Dismissively I said, "Listen, Gordon, it's a long story with an unlikely ending. You really don't want to hear it."

"On the contrary...."

I cut him off short, saying, "Just write up the ticket and I'll be on my way, okay?"

"Sure. Whatever."

My business there concluded, I left the store and drove back over the mountain to my little Shangri La in the pines...right after I stopped at the grocery store for a pound of liver. I was a few hundred dollars poorer and felt like a moron driven by the dictates of lifestyle-induced delusions, but at least I had everything I needed to make my little pond pump solar system completely foolproof. It would all be worth it, I told myself, if things could now just get back to normal.

And, for a time, my wish came true. Things were gratifyingly normal. The two new components worked perfectly, once I managed to get the jumpers and set points on the circuit boards in the right places and calibrated properly. And all five of my pets went back to doing the things dogs and cats are supposed to do, whatever that is. Mick even had the good manners to act surprised when he found liver—grilled, not fried—mixed in with his dog food.

After a few weeks, in fact, I halfway managed to convince myself that Newt had never disappeared into the solar panel, or that I'd had any late night heart-to-heart's with Mick. Or that Wild Willie was really capable of all the devious acts I had accused him of.

Certainly, it was all a dream.

Then I paid another visit to Pickett's Farm & Ranch Supply.

the perfect woman
a perfect path to ... what else?
perfect confusion

Salt. The whole purpose of my trip to Pickett's Farm & Ranch Supply was to buy a couple of 50-pound blocks of the stuff for my calves. The pink kind that has trace minerals mixed in with it. I had no intention of engaging Eli in a conversation about tractor/pond pump wiring— it was all the same to him—or anything else to do with the events of a few weeks before. It was all in the past, now. In fact, I was hoping to get in and out without even setting eyes on the old scavenger.

But, as usual, Fate proved herself to be a playful companion. For, way back by the stacks of barbed wire and welded fence, I saw a sight that made my heart skip a beat. Or two or three.

Eli's grizzled form was the first thing I saw. Instinct, I have found, always draws one's eyes to the most obvious and immediate source of danger. In front of Eli stood a tall fellow in ersatz-faded denims with two women standing beside him. All three had their backs to me. The woman closest to the man had short, dark hair and wore a pair of overalls that would never shrink to fit her slender figure in a million wash-and-dry cycles. The other woman, standing a few paces away, had long, light brown hair, crowning a tall, graceful form that ignited within me a scorching flame of recognition. I knew this woman. It was her; it just had to be. Same figure, same hair, same...; recalling my dream of weeks before, I let my eyes roam to the wall behind her, where I saw exactly what I hoped to see: fan belts. The entire wall was lined with them—some new, some dusty, some tethered to the wall with vast lattices of cobwebs—all hanging haphazardly on nails driven inexpertly into the wall board.

I swallowed hard, even though my throat had gone completely dry, and studied the situation.

She seemed to show only a passing interest in the conversation; obviously she wasn't in the store on any kind of business. Probably just tagging along with her sister or brother.

Suddenly, my attitude toward Eli did a complete one-eighty. Heck, he wasn't such a bad guy. Deep down, I was sure, I really liked the little shiny-headed troll. In fact, I was overcome by an urge to go strike up a conversation with him. Certainly his other customers wouldn't mind my intrusion, would they?

When he saw me walking toward him his eyes narrowed into a menacing glare and his lips tightened into a thin, bloodless line. What a kidder ol' Eli was, pretending like he wasn't happy to see me. His cryptic sense of humor became even more evident when he croaked, "Don't bother, boy. If this has to do with your silly little sun-powered pond pump thingy I can't help you no more." It was only when he planted his hands on his hips and took a step toward me that I realized this was not going to be easy. Maybe not even possible. But as I turned my head to gaze upon the young woman standing a few paces away, I was instantly suffused with an intractable sense of purpose. Eli had better have a gun, if he wanted me to leave. A big one.

I opened my mouth—to say what, I had no idea—when she turned to face me. Her lips were formed into a coy little smile, and a sprightly gleam played in her gorgeous green eyes. My mouth instantly clamped shut, lest I say something stupid. Then she spoke. She could have said anything; a recitation of *row, row, row your boat* would have sounded like the sirens' song, coming from her lips. But she said, simply, "A 'sun-powered' pond pump? Do you mean *solar* powered?"

I wasn't sure I could manage the word "yes," but I gave it a try. She giggled when I chirped it in an octave I didn't know I was capable of.

"Swallow a canary?" she asked.

Eli broke in, saying, "Don't pay him no never mind, young lady. The boy just don't know what he's about, that's all." Good old Eli. I knew I could count on him to smooth over a rocky beginning.

I felt my face growing red, and the inside of my mouth had the taste and feel of sunburned cotton. It was now or never. If I didn't say something intelligent, and fast, I might as well go take a long nap on the bottom of my pond.

"It's a rudimentary system," I heard myself say, at last, "but really quite elegant in its simplicity. Eli, here, got me started with fuses and disconnects, but since then I've added other refinements that make the whole system self-regulating." So far so good. She looked interested, maybe even a little impressed. Should I go for broke? As my left-brain pondered this, my right-brain commandeered my vocal cords, and uttered, "I'd love to show it to you."

Right on cue, Eli butted in. "I'm sure you would, Sonny Boy, but these folks got better things to do than—"

She cut him off like he wasn't even there. "And I'd love to see it," she said, looking at no one but me. "How far away do you live?"

"Not far at all, really."

"Ha! Since when is 25 miles over twisted, unpaved mountain trails 'not far at all, really'?" Eli snipped.

"Twenty," I corrected. "And it's paved most of the way. Sort of."

"Sounds like a real adventure," she said, with an alluring twinkle in her eye. I could have fainted.

Her name was LaVonne. She was there with her sister, Angela, and her brother-in-law, Hank, two electrical engineers who had just abandoned their high-tech jobs in the sprawling, polluted metropolis of Big

City to live on a small acreage near the Wyoming border. They were there buying fencing and other ranching supplies. Things I just happened to know a whole lot about. LaVonne, I learned, was a nature artist, living in their small guest house. I surmised that they would need a fair bit of expert help getting their little operation up and running. And I was just the guy to lend them a hand.

The immediate plan was simple: I'd take LaVonne to see my solar pond pump setup, though for the life of me I couldn't understand why she thought it so interesting. Later we'd all meet at Hank and Angela's to look over their place and see what had to be done to make it workable.

Eli could stay in his store and look for new people to insult.

Things couldn't have been more perfect. So why was I looking over my shoulder for a bolt of lightning?

Maybe it was the way the dogs and cats took to her, once we arrived at my ranch. They flocked to her like she was made of roast beef and catnip. Even Stinky left her throne to nuzzle her way into the act, as LaVonne rested on a tree stump, petting and talking to each one in turn. I'd never seen anything quite like it. Frankly, it was worrisome. What would happen to my fragile new relationship with the Perfect Woman if Mick were to start blabbering

away about electrical potential and extra-dimensionality?

I offered my hand to pull her away from the panting, writhing mass of fur balls and, to my good fortune, she took it. It was like grabbing hold of an angel and I didn't want to let go, but after an enchanted moment of staring deeply into her eyes I knew there would be more opportunities down the road. Better not press my luck.

I then showed her the solar pond pump system I was so proud of. She examined each component in turn, as the water bubbled melodically over the small waterfall I'd built on the shore, and asked several questions about wattage and amp hours and load amps, and so forth. Things I had just recently learned, myself. Though it was clear she knew what she was talking about, the words seemed a little awkward in her mouth, as though she had not long been in the habit of using them.

Fine with me. I'd have rather talked about other things, anyway.

As we walked away from the pond, she asked, "Have any of your animals been acting strangely?"

I stopped dead in my tracks. "Why do you ask?" I wondered aloud, after the wave of surprise subsided.

My hesitation didn't go unnoticed. "Uh huh. I thought so."

"You thought what?"

She turned to face me. I wanted to kiss her, but her stern expression said not now—as opposed, I hoped, to not ever. She said, "Don't you see? It's starting."

The gravity of her voice caught me off-guard. "What's starting?" I asked, in utter confusion.

"Nature is reclaiming what's Hers. It's starting with the animals. I'm sure you've seen subtle signs. Extraordinary intelligence, maybe. Or levels of cooperation that don't seem possible." She paused for a moment, took my hand, and said, "C'mon, admit it. You have noticed, haven't you, Rex?"

I almost said, *just spend a night in my bed when the system's on the fritz and you'll see animal behavior like you never dreamed of before.* But, upon a moment's reflection, I realized just how easily such a statement could be misinterpreted. Instead, I said, "Well, yeah...there has been an instant, or two. I just figured it was my imagination running wild."

She chuckled, and said, "No, it's more than that. A lot more. The truth is, everything is about to flip upside down, and you—we—want to be on the right side of things after the dust settles."

"So it's not the Rolling Stones?" I wondered, more to myself.

Humoring me, she smiled warmly, and replied, "Believe it or not, this is even bigger than Mick Jagger." Still holding my hand, she led me to a bench by the pond. "Care to explain what's been happening around here?"

And so I did. I told her about how Willie had lured Newt into the solar panel—twice—and all about my two late-night conversations with Mick. I spoke of impossible things as though they were an everyday occurrence around the Last Chance Ranch. I spoke until my rigid concept of normalcy melted into a bubbling pool of imponderable probabilities, and when I was all through I fully expected her to tell me that me and my senses had parted ways.

Instead, she said, "Yeah, that's about what I would've expected."

"From me, you mean?" I asked, feeling a little paranoid.

She laughed. "Don't be silly. There's more to this than you think. A lot more."

I waited for her to explain, but she didn't. She just asked me to drive her home. So I did.

As we pulled out onto the county road, I realized I had forgotten to buy my salt. 🌿

*back to the future
a planet begins to change
rodents run the show*

As much as I enjoyed being alone with LaVonne—she was still the Perfect Woman, though the concept had begun to take on a more nebulous morphology—it was almost a relief to be at Hank and

Angela's talking about corner posts and stretcher posts, and the pros and cons of 4-wire, versus 5-wire, fences.

My rancher's brain, I had recently discovered, has a low weirdness threshold, and a lengthy exposition of 19th century animal confinement technology was a warm and fuzzy refuge for it. We walked the perimeter of their 40 acres and they listened intently as I explained the best ways for them to go about shoring up the fences that were already there and building new fences where they were needed. I offered advice for repairing their old outbuildings and the most practical placement of the new ones they had planned. So studiously did they focus on my every word, I felt like a purveyor of some arcane body of knowledge that promised glowing good health over a vastly extended lifetime.

What Hank and Angela lacked in know-how about fence and outbuilding construction, however, they more than made up for with their knowledge about the nuts and bolts of renewable energy. And not just the theoretical side, either—it was everywhere evident they knew how to put their learning into practice.

The wind turbine on the bluff above their house was the most conspicuous example. Looking much like a stripped fuselage with a disproportionately large propeller, I marveled at how furiously the blades spun in the light breeze. "How does it work?" I asked Hank.

He smiled, and said, "Do you want the short version, or the long?"

"How about the short version, for now?"

"Okay, then. I'll give you one word: magnets."

"Magnets?"

"Right," he confirmed. "Three pairs of them, actually, located 120 degrees apart around the inside of the rotor. And each pair of opposing polarity."

"You mean like north and south?" I asked, remembering my high school science classes.

"Well, yes, but 'positive' and 'negative' work better when you're talking about electricity."

"I suppose," I agreed.

"Anyway, when the propeller turns, the magnets revolve around the stator, and induce three separate—but 'in phase'—currents within the stator's windings."

"Stator? What's that?" I asked, feeling very much out of my element.

"It's like a cylinder with coils of wire wrapped around it length-wise. Three primary coils, in this case," he explained.

I scratched my head. "Pardon my ignorance, but how does that make electricity?"

He grinned broadly, and answered, "The accepted term is 'inductance'. As for how or why it works, no one knows, exactly. All we can say for sure is that a magnetic field will cause a current to flow in a wire, just like a wire carrying current will produce a magnetic field around itself. So when you spin two oppositely charged magnets around a coil of wire it 'induces' an alternating electric current within the wire. It's kind of spooky, when you think about it."

"Alternating current?" I asked, remembering my last conversation with Mick. "You mean like house current?"

"Well, sort of. House current is single phase, not three-phase. Also the wind generator voltage is lower. And the frequency of the current varies with the wind speed."

"Frequency?"

"Right. The number of times per second the current alternates—or cycles—from positive to negative. Normal house current is regulated at 60 hertz, or 60 cycles per second."

It was beginning to sound complicated, which meant I was starting to feel stupid. Groping for familiar ground, I remembered my pond pump setup, and asked, "So you run the current through a charge controller, and then into your battery bank?"

"Exactly," he confirmed, "except it's not like the charge controller for your little 12-volt DC solar module. This one has rectifiers to change the AC into DC, so it can be stored in the batteries."

"You can't store AC current?"

He laughed good-naturedly, and answered, "No. It would run out again as fast as you put it in. AC is a pulsing current. It's defined by its motion. Once you stop it, it's not AC anymore."

Changing the subject, I asked, "So how much power does your turbine produce?"

"It's rated at 1,000 watts when the wind blows at 32 mph or so, but it'll go higher for short periods."

"And what will 1,000 watts run?"

"Oh, a dishwasher or hair dryer, or maybe a circular saw."

"That's all?" I said, a little disappointed.

"That's enough," he answered, "because you don't usually run heavy loads often, or for a long period of time. And the power you produce when you're not running those things goes right into the batteries. Anyway, our solar array is the real meat and potatoes of the system. It's rated at just under 2,000 watts."

He pointed to two large aluminum frames in front of the house mounted on heavy poles set in concrete. Each of the frames held eight solar panels—or modules, as Hank called them, being an ex-geek and all—and every one was a lot bigger than the little one I had for my pond pump system. I felt an annoying twinge of solar panel envy.

I said, "So let me get this straight: the wind generator produces AC current that gets changed into DC and stored in the batteries. The solar panels produce DC that's likewise stored in the batteries. But the house runs on AC. How does that happen?"

"A nifty thing called an inverter," he said, leading me down the stairs into the basement. A large, white metal case was mounted on the wall amidst a lot of other smaller boxes with displays and switches. I was pretty sure one of them was the charge controller for the solar array; the other stuff didn't look in the least familiar. A pervasive hum filled the room; it sounded like power waiting to happen. Hank pointed to the largest component—the one with the biggest LCD display and the most buttons—and said, proudly, "The inverter. It takes low voltage DC current from the batteries here,"—he pointed to a large box on the floor beside the inverter—"and converts it into usable 120-volt AC house current, pulsing at 60 hertz."

"How does it do that?" I asked, truly curious.

Again, he asked, "Long version, or short?"

"Short," I replied.

"Okay, then. One word: magic."

"Sounds good to me. I'm beginning to like the mystery."

Hank said, "Good. Let's go eat."

When I first arrived, I'd noticed a small building several yards from the house. By the way the dirt was disturbed around the foundation, it was clearly of recent construction. It had its own solar array, and several vent stacks poking through the roof. At last, my curiosity forced me to ask: "Hank, what in the world do you have in there?"

He flashed a grin, pointed to the building and said, "There? Oh, you're not quite ready for that, yet." Then he turned and led me into the house.

To that point it had been a fairly normal day. I'd learned a lot of technical stuff, and had even been able to share some of my knowledge—antiquated though it was—about ranching, fencing and barn building. It was only after we sat down for dinner that the conversation took a disturbing detour into uncharted territory. It began, innocently enough, when Hank asked me if I'd noticed more power outages in the last year or so. As a matter of fact I had, though I hadn't given it much thought. I just assumed it was decades-old transformers, lines and power poles finally wearing out or rotting in the ground.

Hank shook his head. "Squirrels," he said.

"Squirrels?" I repeated.

"And beavers," Angela added. "They chew down the poles while the squirrels nibble on the wires and short out the transformers."

"Don't forget the prairie dogs," LaVonne piped in, turning her gaze to me. "They chew the insulation off buried lines."

Prairie dogs I could believe. They were born troublemakers. But squirrels and beavers? I asked, "I'm sure there's a point to all this sabotage...?"

Angela fielded the question. "The animals have decided things are starting to get out of hand. We're using up natural resources like there's no tomorrow, just to manufacture and power a bunch of junk we don't really need. The animals all accepted it, back when we all lived simpler lives and coal-burning power plants were the apex of technology. But now things are different. We have clean, viable options. Solar and wind power, for starters. Even micro-hydro power. There is no need for the planet to suffer further insult at the hands of the energy moguls. The animals are just trying to give us a little push in the right direction."

"And if we don't take the hint?" I asked.

"Then believe me, things could get really nasty."

LaVonne walked me to my pickup when it came time for me to leave. "Interesting family," I told her, after I was sure we were out of earshot of the house.

"But a little nuts, maybe?"

"I didn't say that," I protested.

She offered one of her ice-melting smiles, and said, "You didn't have to. I'd think we were nuts too, if I didn't know better."

"You mean you really believe all this stuff about squirrels and beavers?"

She laughed, and said, "That's a strange question, coming from a man who has late-night chats with his dog."

I'd momentarily forgotten I'd spilled the beans on that one, and felt like I'd been chastised when she reminded me. "Yeah, but still....." I mumbled, lamely.

"Don't worry—you'll come around." With that, she kissed me on the cheek, said, "See ya!" and followed the path back into the house.

As I watched her disappear behind the door, I began to seriously wonder what I had gotten myself into, simply because I wanted a novel little solar-powered pump for my pond. It had already elicited mind-warping behavior from my animals, cost me hundreds of dollars more than I had planned to spend, and indirectly led me to the Perfect Woman, who then makes me privy to a vast wildlife conspiracy to drive the electrical utilities out of business.

The thing really should have come with a stronger warning label, I concluded, turning my pickup in the general direction of home.

As I left the high plains of no-man's land of far northern Colorado and eased into the first mountain valley on the way to my ranch, it occurred to me that either LaVonne was not perfect, or she was telling the truth, since logic dictated that the Perfect Woman would not be able to tell a lie. Unwilling to entertain the notion that she was anything other than what I believed her to be, I was forced to conclude she knew what she was talking about. There had been a lot more power outages lately, hadn't there? And if small, furry animals were the culprits, Planet Power Corporation would want to keep it quiet, rather than admit to being bested by a marauding band of buck-toothed varmints. But what about the press? Wouldn't they jump on a story like this? Maybe not. They loved prairie dogs and their ilk. If they blew the whistle on what all these animals were really up to there'd be public outrage and wholesale slaughter. The streets and alleys would fill with gun-toting vigilantes, determined not to miss Monday Night Football because of some agenda-driven gopher gang.

If it truly was a conspiracy, it was a beautiful one. Everyone would blame Planet Power for inefficient service and, to make matters worse, they'd be forced to raise rates to cover their extra costs, adding more fuel to the public opinion fires. And no one could—or would—say a word about what was really happening.

I laughed to myself as I framed a mental picture of Gilbert Gigabucks, the CEO of Planet Power Corporation, holding a press conference. With wild eyes and sweat running down his forehead in rivulets, he'd exclaim, "It's squirrels, I tell you! Squirrels and beavers and prairie dogs! Oh, my! They're the reason for all the blackouts and the 200 percent rise in your electric bill, so blame them, for crying out loud, not me!" They'd lock him in a room with vulcanized walls and throw away the key.

I shook my head. It was crazy. So why was I beginning to believe it? Because the world was a crazy place, and getting crazier by the day, I told myself. Besides, I was enamored with the woman who told me about it; I refused to believe she was off her beam.

Then, just as I crested a small rise on the winding road leading to the ranch turnoff, I saw the confirmation I needed. As my headlights pointed downward they illuminated a pair of beavers at the base of a power pole. I worked my brakes, slowed to a stop in the middle of the deserted road, and watched.

They were not in the least disturbed by my presence. In fact, they both seemed to be smiling (though, admittedly, I would be hard pressed to qualify the difference between a smiling beaver, and a frowning one). I could see the pole was almost chewed through; only the overhead lines were keeping it erect. "What're you gonna do now, little guys?" I whispered to myself, amused.

I didn't have to wait long for an answer, for out of the shadows emerged an enormous bull moose. He lowered his head while one of the beavers chittered something in his ear, then stepped up to the pole and straddled it with his antlers. The pole weaved back and forth as the bull pushed and relaxed, pushed and relaxed, until, with a great snap and a shower of sparks, the pole crashed down amidst a tangle of wires. The beavers jumped up and down, thumping their broad, flat tails against the ground and clapping their little paws with delight, while the moose trumpeted a triumphant refrain.

Awe-struck by what I had just seen, and confident now that LaVonne wasn't crazy, I slowly continued my journey home.

My giddiness at the sight I'd just witnessed was quickly damp-ened, however, when I realized my house was enshrouded in darkness. The pesky beavers had cut my power. ✺

a bright day dawning
a new order now begins
drive on, red Corvette

I could hardly believe my eyes when I found LaVonne standing beside my bed the next morning. She offered me a bright smile and a steaming cup of coffee as I sat up and wiped the sleep out of my eyes. "Morning, Sunshine," she said, melodiously. "Let's get moving. Things are happening fast."

Once I realized she really was there in my bedroom, I had several questions, starting with: how did she get in the door? (probably didn't bother to lock it, as usual), and what time did she have to get up to drive all the winding back roads between her place and mine, and still find me in bed? And what time was it, anyway?

"A little past nine. How do you make a living, if you sleep all day?" she chided.

"Sometimes I wonder that myself," I croaked, taking a sip of coffee. It was chewy. "What the—?"

"Power's out. Had to make it on the gas stove. You should really keep some instant around, don't you think?"

"I'll be sure to put it on my list," I promised.

Not wanting to risk a shower with the well out of commission, I dressed, threw a little water in my face, and combed my hair. By the time I made it down the stairs she had already fed the dogs and cats and horses; I didn't even want to know how she managed it—sometimes it's easier just to believe in magic.

Sliding myself into the passenger seat of her immaculately restored, candy apple red '64 Corvette convertible, I told myself that she was most likely an angel, and wherever she took me would be for my own good. It was a soothing—albeit thoughtless—philosophy.

Our first stop was the scene of last night's crime. There was quite a commotion along side the road. The destroyed pole rested on the back of a big Planet Power semi rig. A pair of workmen tamped the ground around the new pole, while another pair re-stretched the overhead wires from a cherry picker mounted on a smaller truck. A staunch-looking man in a white shirt and tie stood on the side of the road with his arms crossed, supervising the repair operation.

LaVonne stopped by the men at the base of the pole, and asked, "What happened here?"

One man kept working, the other looked up with a nervous smile. He started to say something, then stopped and glanced at the dour-faced executive on the shoulder of the road. "Logging truck must'a sideswiped the pole sometime in the night," he lied.

I replied, "Yeah. You really gotta watch out for those buck-toothed loggers."

He gave a shaky laugh, and answered, "Ain't that the truth?"

"Yeah, ain't it though," LaVonne snipped, giving the suit an icy stare before popping the clutch and laying a squealing, smoking patch of rubber that sent the poor man scurrying for the side of the road.

"Good lord! What do you run in this thing? Nitro?" I asked, after the screeching subsided and I was able to peel my back from the seat.

She shook her head. "Hydrogen."

"Hydrogen?" I said, in disbelief. "I thought that technology still had a lot of bugs in it."

"It does," she replied, never taking her eyes from the winding road speeding beneath the car.

"Okay then, how....?"

"I just happen to know a good exterminator," she answered, gracing me with an enigmatic smile as she slid a tape in the tape player. I started to ask her what she meant, but was suddenly drowned out by Mick Jagger hammering *Wild Horses* through the 'Vette's speakers.

What a woman.

As we sped down the road in her pollution-free hot rod, it occurred to me that there was a lot to this renewable energy business that I didn't understand. Hank's whirlwind tour helped, but I was still more confused than enlightened. I turned down the music and said, "You know, all of this solar and wind stuff is great, but I still don't understand how it all works."

She smiled, and said, "Maybe a little analogy will help."

"Couldn't hurt," I agreed.

Thinking, she stared into the distance to the horizon beyond the road. After a moment she said, "Okay. I've got it." Glancing in my direction, she said, "You've got your little pond pump system pretty well figured out, right?"

"Yeah."

"Good. We'll take it to the extreme, then. Let's say, for instance, that you decide you want the pump to run at night, hoping that perhaps the sound of running water will attract wildlife to your pond."

"That'd be nice," I agreed.

"Okay then. All you need to do is install another battery and a second solar panel, making sure that the combined panels—wired in parallel, of course—are powerful enough to charge the batteries for nighttime use, while still running the pump during the day. As long as the sun shines every day, your simple little system works great."

"Okay," I interrupted. "You said 'wired in parallel'. I remember Hank talking about parallel and series wiring. Some things he did one way, some things another. What's the difference?"

Not taking her eyes from the winding road, she answered, "When you wire something in parallel you increase the amperage, but not the voltage. That's because there are multiple paths for the current, so it doesn't increase the 'electrical pressure' in the system. But a series circuit only allows one path. This increases the voltage because it does raise the electrical pressure."

"Got it, I think."

"Good. Now back to the analogy...You're finally happy. At least until, after a long stretch of cloudy, windy weather, you notice the pump is sitting idle, your batteries are practically drained, and the wildlife have disappeared..."

At that point I stopped her and remarked, "You forgot to mention the part about my rotten yellow cat luring Newt into the solar panel."

"It's not a necessary element to the story," she answered with teacher-like seriousness, before continuing. "...You get out your spotting scope and train it across the valley, on your neighbor's place. There you spy *your* deer frolicking around *his* pond, the one run by the big pump that's hooked into the Planet Power Corporation grid. You don't want to invest in the extra solar modules and batteries to keep the pump running day and night on such sparse sunlight. Instead, you decide to add a wind turbine to the system to supply the pump with power on windy days when the sun goes into hiding."

Again I interrupted, asking, "How'd you learn all this stuff? I thought you were an artist."

"The art is from my mother's side of the family," she explained. "My father comes from a long line of engineers, machinists and tinkerers." She paused, glanced in my direction to see if I had anything more to say, then continued. "It's a fairly large wind generator that sends AC current through the wires, rather than DC. Luckily, it comes with its own charge controller—one with rectifiers that convert the AC to DC before sending the current to the batteries, since—as you now know—batteries can't store alternating current."

"Just like you can't catch a moonbeam in your hand?" I had to ask, just to see her gorgeous eyes flash. They did.

"Finally, you're completely satisfied," she went on. "At last you have a pond pump that works under any meteorological condition. But after months of watching a feeble column of water dribble over a few rocks, you decide your little 12-volt pump just doesn't provide enough action. You check around to see what's available and after a little research you find just the pump you want. But to your dismay, this pump runs on 120-volt AC, and a lot of it."

"Sounds like things are about to get expensive," I predicted.

"It doesn't matter. You're a stubborn Irishman. This is *the* pump, so you upgrade to a bigger wind generator and buy more solar panels. And, of course, you add more batteries. Now you've got the generation and storage capacities you need to run the new pump, but not the right kind of current. One more component should do the trick."

"I'll bet this is the part where we add the inverter," I said, hopefully.

"You got it, Cowboy," she smiled. "You add a power inverter between— where else?—the batteries and the pump. This gives you the steady, clean, AC power you need to force a powerful spray of water out the blow- hole of a three-foot- high whale with a mermaid sitting on its back..."

"I love this thing! Where can I get one?" I exclaimed.

"I knew you would. Unfortunately, it only exists in my imagination."

"Hey! Wait a minute..." I began to protest, just as she pulled into a parking lot and killed the engine.

It came as no surprise that she had driven me to the Super Solar Megastore, where I'd bought my charge controllers, way back when all I wanted was normal pets. LaVonne turned to me, and said, "Okay, it's decision time."

"I suppose," I answered.

"You've seen what's happening with your own eyes."

"Uh huh."

"And I'm sure you can imagine how bad it's going to get, before things settle down again," she added.

I could. Rodents and ruminants were just the beginning, I was sure. What would happen when birds started shorting out overhead lines and fire ants began crawling inside all of Planet Power Corporation's delicate equipment? It would be the end of fossil-fuel-burning power plants. Which meant the stuff they sold in this store would cost ten times what it did now, if it was available. Which it wouldn't be.

She took my hands in hers, then, looking me straight in the eye, asked, "So what's it going to be, Cowboy? Off the grid now, while you still have a choice, or later, when you don't?"

"Let's do it." I told her, losing myself in her alluring green eyes.

"Let's," she agreed, letting go of my hands to fish a list from her purse. She said, "I had Hank work this up last night. It's probably not everything you're going to need, but it's a good start."

I took the list and looked it over. A lot of the stuff I recognized from my experience with the pond pump system; other things—like the DC disconnect, auto-transformer and combiner box—I vaguely remembered from my conversations with Hank. I could sense that I would be ascending a sharp learning curve in the very near future.

We walked through the doors and LaVonne handed the list to Gordon. He recognized me right away, and also seemed to know LaVonne. You could see it was driving him nuts trying to figure out what we were doing together. Good. Let him wonder.

It took him a few minutes to work up the bid. After he was through punching numbers he hesitated, no doubt wondering how much discount I would expect on such a large order. Being a seasoned horse trader, I wasn't about to show my hand until he showed his. Finally, he handed me the total.

I had my reaction planned before even looking at bottom line. "You can't be serious!" I protested. "We can get all this stuff a lot cheaper off the Internet, even paying for shipping!"

His mouth drew into a knot of indignation as he studied me to see if I was bluffing, then wriggled into a smile when he mistakenly concluded I wasn't. "Okay, let me see what I can do."

When he handed us back the revised total, LaVonne took over the negotiations. With a voice that could turn battery acid into honey, she said, "C'mon, Gordon... this isn't your best price, now is it? We know you can do better."

"Well...."

"And Gordon," she added, just as sweetly, "can we expect prompt—and free—delivery?"

The final total was most agreeable; still, I felt like I'd been broadsided. How much livestock would I have to sell to pay for all this stuff?

"Don't worry about it," LaVonne said, sensing my concern. "Of all the alternatives, this one is the cheapest."

As things turned out, she was right. 🔥

 hard work and plenty
disaster's seeds have rooted
the cat bides his time

Never is a rancher's mind so content as when it's filled with a sense of purpose. Like any other vessel, however, a mind can only hold so much of anything before it begins to spill over the sides and turn sour. So, after a week of digging holes in rocky ground and setting the bases for the wind tower and solar array in concrete; digging trenches for the underground wires; building a box to hold the batteries, and drilling holes through the side of my house for all the conduit-housed wires running into it, I'd had more than my share of purposeful contentment. I was ready for some good, old-fashioned laziness.

To my dismay, however, it was not to be, for time was becoming an increasingly sparse commodity. The scene I had witnessed on the side of the road a week before was repeated again and again around the area—and, from what I could gather, around the country—though neither the media nor Planet Power would fess up to the real cause of the ever-more-frequent power outages.

The one crack in the carapace of this far-reaching cover-up conspiracy came in the form of a widely-broadcast home videotape of a thousand or so plump crows sitting on a power line across the road from a distraught suburbanite in northern Iowa. Before the viewer's eyes, the crows began to lift from the wire and then settle back down again, in a purposeful, wavelike motion. And not just once. They did it until an undulating resonance whipped along the wire from one pole to the other, and back again. After several minutes of this violent, serpentine motion, the wire snapped from the pole and the crows all flew away, caw-cawing in what could only be delight. The media, not wanting to admit that crows could be so incredibly smart—and no doubt fearing that the nation's crow population would be immediately and severely stressed by a million shotgun-toting farmers—laughed it off as an inexplicable freak of nature. LaVonne and I knew better.

Though there were frequent outages all around me, my ranch remained unaffected. Somehow, the forest creatures knew what we were up to and had no intention of doing anything to slow our progress—a fact for which we were most grateful.

Whenever I worked alone, squirrels watched me from the treetops, and scores of birds scratched the ground around me, pretending to search for seeds. They were sentries, of course; eye-witness reporters from whom originated communiqués that traveled throughout the animal kingdom with a speed and efficiency that surpassed contemporary satellite communication. I marveled at the possibility of bamboo-chewing pandas in China and gazelle-stalking lions in Africa's Serengeti nodding approval at my endeavors to free myself from Planet Power Corporation's smog-ridden stranglehold on the world's electricity.

I assumed the animal grapevine was a sound-driven network—akin to the strange and beautiful sonar communication used by whales and dolphins—but LaVonne assured me it was telepathic in nature. Pure thought—unsullied by words, she claimed—was being instantaneously broadcast into the ether by birds and mammals, and even reptiles and insects, all around the globe. They knew; they all knew.

Bizarre as it seemed, it was hard to argue with her. When she found the time to come and help me with the innumerable tasks at hand, the squirrels came out of the trees and surrounded her like a large litter of attention-starved puppies. Competing species of birds would momentarily put aside their petty differences and alight upon her shoulders. It was a wonder to see.

Even the dogs made an uneasy peace with the not-quite-wild animals that hovered around LaVonne. Mick and Amy lay down beside her, grudgingly allowing the squirrels and chipmunks to use them for perches. Newt, who was genetically hardwired to chase any small thing that moved, kept her distance—unable to endure the insult of having a chipmunk or two, sitting on top of her head.

The cats were a different matter. Stinky watched the forest creatures with unwavering anticipation, her eyes following their every movement. To her credit, I believe it was more from fascination than any predatory designs, since she never once made a threatening move. Then there was Willie, the cat for whose demented pleasures the entire world existed. The telepathic waves of goodwill that were sufficient to transport Amy and Mick to Nirvana, and at least keep Stinky and Newt at bay, just irritated Willie, like chaff in the wind. From different vantage points he launched one attack after another, each one repelled, either by an army of squirrels, a squadron of birds, or simply a firm hand latching onto the scruff of his neck.

Finally, pecked and scratched, angry and frustrated, the big yellow cat retreated to the shadows and brooded darkly. It was worrisome.

Per Hank's instructions, I had run all the wires and installed all the components. The only task that remained was that of hooking everything together and flipping a switch or two. Knowing how tricky it could be—and how ignorant I was of the whole process—Hank came by early one day to assist me. He worked from early morning to late afternoon, while I mostly watched. When there was nothing left to do but wire the leads from the AC power inverter into the main breaker panel, Hank asked, "Are you really ready to free yourself from Planet Power's tentacles?"

I was.

He solemnly handed me a pair of wire nips and pointed to the electrical meter on the utility pole closest to the house. "Then do it," he said.

With neither pomp nor fanfare, I first cut the wire with the leaden seal that held the meter's flange in place, knowing full-well that I had to be breaking some sort of law. I then took a deep breath and prepared to yank the meter from its receptacle on the pole. As my nervous fingers wrapped around the ominous-looking glass jar that housed the meter's inner workings, a score of squirrels, chipmunks and birds formed a circle around me. A pair of bull elk emerged from the trees to watch. I gave the meter a hard pull and it came free, to a chorus of chirps and chitters and throaty refrains. The elk bugled in tandem, their eerie calls echoing across the distant hillsides.

Electrically speaking, the house was now dead—though not for long. Quickly Hank went to work to resuscitate it. Within 20 minutes the house was running again—this time on clean, free, non-depleting (this was a cool new term I'd recently learned) energy from the sun and wind. Hank shook my hand and LaVonne rewarded me with a long, satisfying kiss. I was elated. I felt like I had just crossed a vast chasm between the old form of humankind, to the new.

I took LaVonne in my arms and returned her kiss. She didn't object, even though my brazenness turned her brother-in-law's face a deep shade of red. "Why, Hank," she said, with a chuckle, "I do believe we've embarrassed you!"

He answered, "Me? No...it's, uh...just a hot day, is all. Anyway, I think it's time for me to go." He gave me several pointers on potential bugs that might creep into the system until it was fine-tuned, and how to deal with them. Then he drove away.

"Dinner in town?" I said to LaVonne, after Hank was gone.

"You buy, I'll drive," came the reply.

It sounded like a fair deal to me. As the sun began to set and LaVonne and I drove away, happy and carefree, it seemed as though all the world was finally at peace. All except for a spurned and sulking yellow cat that crawled out from hiding to ponder deeds most unsavory. 🔥

an antithesis
yellow cat has learned his code
darkness delivers

As we sped into town in LaVonne's hydrogen-powered muscle-car, it occurred to me that perhaps not all of nature's creatures would look forward to the budding energy revolution with the same hopeful anticipation as the winged and antlered and chisel-toothed vigilantes. Some, those who thrived in a polluted world, might feel severely threatened by any philosophy that required its followers to live as though the planet actually mattered. If people suddenly began to take responsibility for the energy they used and the means they employed to produce it, might they not also become mindful of other wasteful practices, such as the millennia-old tendency to pile garbage in big, smelly heaps? If the energy revolution ever caught on, it would mean a wholesale restructuring of the refuse chain. Everything would be recycled. People's homes and yards would be neat and efficient, city streets would be far cleaner, and trash dumps would become a thing of the past. What, then, would become of the unnumbered populations of vermin—rats and cockroaches, for instance—if humanity were suddenly to treat the world with same respect as other animals? Might they not resent the change, and try to sabotage our efforts?

Unable to shake the feeling that this was more than my imagination, I shared my concerns with LaVonne once we were seated at a small, candlelit table in the one of Big City's coziest and least known restaurants. She just smiled, and said, "You're catching on."

"Catching on to what?" I had to ask.

"The way of the world," she answered. "Of course there's going to be a backlash. It's just the way things are. Remember Newton, and your high school physics? For every action there's an equal and opposite reaction?"

"But we're talking about animals here, not rockets," I pointed out.

Undaunted, she replied, "Okay, then…how about yin and yang? Or Hegel's dialectic? For every thesis there's an antithesis? Eventually the two resolve into a synthesis?

"Yeah," I jabbed, "Ol' Karl Marx tried putting Hegel's dialectic into practice. Last I heard, it wasn't working out too good."

She brushed off my objection with a wave of her hand. "Marx's idealism stifled the course of societal evolution. He didn't understand Hegel at all."

"Does anyone?" I wondered.

"Good point," she conceded. "Just the same, it's nothing you need to worry about. Remember the simple little truism, 'whatever must be, will be'? These things have a way of working themselves out." Picking up her menu, she said, "Anyway, are we going to order some food, or should we just sit here all night and discuss 19th century German philosophy?"

I thought she was making light of what might end up being a real problem, but she didn't seem to be the least bit worried. Maybe she was just playing down the issue because she didn't have a solution. Or maybe she knew something I didn't. Either way, she had nothing left to say on the subject, so what difference did it make? There wasn't much either of us could do about it at the moment. Besides, it was a pivotal day and a good time for a little indulgence. We ordered a bottle of Colorado Merlot and enjoyed it over a perfectly roasted chateaubriand. We discussed many things of scant consequence and enjoyed each other's company immensely.

I tried my best to steer clear of "shop talk," feeling the night was better suited for more intimate subjects, but after awhile our conversation took the inevitable turn in that direction. "You know," I said, finally, "I've just cut my ties with Planet Power as though it were a well-planned move, but the fact is I know basically nothing about how my place is powered now. It's a little unnerving."

"What did you know about how it was powered before?" she asked.

"What was there to know? Juice came through the lines, everything worked most of the time, and I paid an electric bill every month."

She laughed, and said, "Well, now things will work all the time, and there's no more electric bill to pay."

"Yeah, but..."

"Maybe I should finish the little analogy I started the other day on our way to town?"

"Why not?" I agreed. "You were just getting to the really cool part about the mermaid and the whale."

"Right, okay, let's see...I've got it." She took a breath, and began. "At last, you are profoundly satisfied with your little system. So popular is your pond, now that the pump runs day and night, an entire wildlife ecosystem has blossomed around it. You only have to look out the window to feel you are on safari. Then your luck gets even better, for as fate would have it, you chance to meet the Perfect Woman..."

"The 'Perfect Woman'? LaVonne, how did you—?"

With a knowing smile, she said, "Because I'm a woman, you starry-eyed cowboy. Now pipe down and listen."

I did, and she continued: "Matrimony follows a whirlwind courtship. All is right with the world..."

"This is really getting deep," I told her, my heart fibrillating wildly.

"It gets deeper," she assured me, before returning to her tale: "...But now your new bride delivers a crushing observation: she has always thought that your anatomically-correct bronze mermaid—perched on the back of a smiling whale—is, well...voyeuristic."

"Really?"

"Uh huh," she answered, nodding her head.

"Dang!"

"...Wouldn't it be better, she innocently asks, if you enlarged the pond—an acre would be about right, she thinks—then built a 20-foot high rock-and-mortar mountain in the middle, from which would gush a miniature Niagara of bubbling water?"

"Now wait a minute here!"

She smiled and flashed me a don't-interrupt-me look. "She's right, of course—you said yourself she was perfect. So, while the excavating crew enlarges the pond to the size of a small lake, and the stone masons build a re-creation of the Matterhorn in the middle of it, you buy another wind generator, a vast solar array, and several more batteries, because you know the old pump for the mermaid and the whale just isn't going to cut it anymore.

"The new pump, as you might've guessed, runs on 240-volt AC, and your inverter only produces 120-volt AC. Should you buy a new inverter, you wonder, and wire it in series to the old one to double the voltage, or is there another way? You do a few calculations and discover the original inverter produces more than enough continuous wattage to drive the new pump, so you take the much cheaper route of installing a 240-volt transformer between the pump, and the inverter."

At that point, I interjected, "It doesn't matter what it costs, Honey, because I had to sell my last horse to pay for the mermaid you just carted off."

"So why don't you pretend you're rich and let me finish the story?"

"But every time I pretend I'm rich, I end up poor, again!" I protested.

Ignoring me, she continued: "The transformer works perfectly, and your waterfall gurgles to life. It is truly a sight to behold. At least until darkness sets in, and you have to view it with a flashlight. So, with a gleam in her eye, your new bride asks if it would be too much trouble to put in a few teensy-weensy lights around the waterfall, maybe twenty, or so? Sure, no problem, you say. But low-wattage compact fluorescent bulbs, you quickly realize, don't run on 240 volts. To remedy the problem, you simply install a 120-volt breaker panel between the inverter and the transformer, to provide a few 120-volt circuits before the current changes to 240 volts to power the waterfall."

"I'm pretty clever, aren't I?" I just had to say.

"Maybe too clever, by half?" she warned, before ending the analogy. "At last: your wife is elated, your waterfall is the eighth wonder of the world, your lake is brimming with fish and exotic waterfowl, and you are eminently proud of yourself for putting together an unassailable solar and wind electrical system." Finished, she took a deep breath and beamed me a satisfied smile.

"Wow," I exclaimed, "this is too much to think about all at once." And I wasn't talking about her lengthy exposition on wind and solar power.

She reached across the table and took my hand. "Don't act so surprised, Cowboy. I knew you were the Perfect Man the minute I set eyes on you."

"Did you ever...I mean, did you have...?"

"Dreams about you?" she asked, finishing my thought. "Sure. I knew just what you would look like before I ever saw you."

"Me too," I said. "About you, I mean."

I was dumbfounded; couldn't think of a thing to say. So, I asked, "Uh, this pond pump setup you just described...do you really want me to build this thing?"

She laughed. "No, silly! It's an analogy. No one in their right mind would spend that much money on a pond pump system. It was meant to help you understand the system you just installed. As a matter of illustration, I took the pond pump idea to a ridiculous extreme."

"That's an understatement," I murmured, then asked, "But how about the mermaid?"

"We'll talk about it," she promised. "But first things first. Let's go back to your place."

On the way out of the restaurant, I asked, "By the way, did you enjoy dinner?"

She offered a sympathetic smile, and replied, "Yes, it was delicious. But..."

"But what?"

"To tell you the truth, I'd have rather had green chili burritos and a bottle of Mexican beer."

No doubt about it—I was in love.

I was hoping the conversation on the way home would wander back into more intimate territory, but LaVonne wasn't one to stoke a hot fire. Instead, the engineering side of her genetic profile kicked in, and she asked, "Okay, let's see how much you've learned. Can you explain the difference between voltage and amperage?"

"Sure. Voltage is the electrical potential of the system, while amperage is the rate of flow."

She furrowed her brow, and said, "That's a copout. It's the same confusing explanation they give to high school kids to make sure they never actually understand this stuff."

Unruffled, I answered, "How about I explain the relationship between the two?"

"Go for it."

Maybe I didn't entirely understand all the everyday, practical aspects of my new system, but I'd thought long and hard about voltage and amperage ever since my first talk with Mick. I was ready to show off a little. I said, "Alright, try this: voltage is to amperage what gravity is to matter."

That earned me a smile. She said, "That's pretty deep; almost poetic. But what does it mean?"

I thought for a minute, then said, "Let's say you've got two cliffs. One is really high, with a small rock resting on the edge. The other is quite a bit lower, with a much larger rock on the edge. You give both rocks a little push. If you've chosen your rocks and your cliffs carefully, you'll discover that both rocks hit the ground with the same force."

"That's neat," she said. "Let's see, now; the two rocks must be analogous to loads of different amperage, and the cliffs of different heights would represent different voltages, or electrical potentials, right?"

"Uh huh. And so the force with which they hit the ground would be...?"

"Wattage, of course."

"You've got it!" I told her.

"But wait a minute," she objected, "Once the rocks have come to rest the big one still weighs more than the little one."

"True," I agreed, "but unless you've got another cliff nearby to roll if off of, it really doesn't have any more potential." I could see she wasn't quite satisfied with this answer, so I said, "Okay, forget the rocks; let's try water falling through a pipe, like in a hydroelectric dam. It works better, anyway."

"You don't say?"

"Oh, but I do. Consider this: the force exerted by water flowing through a pipe at 5 gallons per minute, with a 100-foot vertical drop, is the same as 100 gallons per minute through a pipe with a 5-foot vertical drop. In the same way, 5 amps of current being pushed through a wire by 100 volts provides the same wattage as 100 amps of current at 5 volts. In either system the variables are interchangeable."

"Really!" she exclaimed. "And all this from a cowboy who a week ago didn't hardly know a watt from an amp!" She thought for moment, then said, "But you forgot to mention that in both cases the pipe, or the wire, needs to be properly sized so you don't lose too much to resistance."

"Good point," I agreed. "In fact, your observation helps to round out the analogy. What happens to a pipe's carrying capacity when you double the diameter?" I asked.

She replied, "Well, you quadruple the surface area of the cross-section, so you can push four times the amount of water through it."

"Exactly. Now, what happens when you double the voltage of a system, without changing the overall wattage?"

She rested her chin between her thumb and forefinger, then said, "Let's see...if the wattage remains the same, then the amperage is halved when the voltage is doubled. If I remember right, it means the wire will have four times the carrying capacity. Right?"

"I rest my case."

Narrowing her eyes, she asked, "Did Hank teach you all this stuff, or are you just that clever?"

"LaVonne! I'm shocked!" I exclaimed, a little pretentiously.

"He did, didn't he?" She asked, poking me in the ribs.

"Some of it," I admitted. "But mostly Hank likes to talk on a more philosophical level. For instance, he's fond of pointing out that no one really understands why it all works the way it does."

"So how come we have all these formulas to perfectly describe electrical behavior?" she asked.

"Oh, we know how it works, well enough. Just not why."

"Give me a for-instance."

"Okay, take inductance. As Hank explains it, we all know that an electrical current running through a wire will induce a magnetic field around the wire. Or vice versa. It's the basic principle behind transformers, motors and generators, and even radios and wireless communications. But no one knows why it works the way it does."

"Interesting."

"Oh, that's just the beginning," I told her, as she turned off the county road onto the drive leading to my house. "Hank ties it all in with theoretical physics. By the time he's finished he's got you believing that an electrical current and a ham sandwich are just different manifestations of the same thing. Then he smiles and says, 'now why do you suppose the world works like that?'"

"Sounds like Hank, alright," she agreed. "But look at it this way: if he's right—about the ham sandwich thing, I mean—then it would go a long way toward explaining how Newt keeps getting in and out of the system."

I hardly heard what she said. "We've got problems," I announced, looking toward the house.

The single light I had left burning in my office was blinking on and off at a seemingly irregular rate. As I studied it, however, I realized the spacing and duration of the blinks was cyclic. Remembering my former training as a ham radio operator, I quickly recognized what it was. I felt an icy shiver run down my spine.

LaVonne was perplexed. "What is it?" she asked.

"Morse Code," I answered.

"Oh, no! What is he saying?"

"PACKRATS...UV...THE...WURLD...UNITE."

"Willie?" she asked.

"Who else?" I sputtered, through clenched teeth. 🌿

 a packrat party
a lofty pursuit, en masse
Too smart, yellow cat

The scene before us as we pulled into the yard was like something out of a John Carpenter film. Packrats were pouring out of the hay barn and scrap piles, the equipment shed and wood piles in droves, their little black beady eyes glistening in the headlights as they scurried toward the house. The three dogs and Stinky, the cat, had tried to set up a defensive perimeter around the house, but there were too many rats for the four of them to fend off; whenever one of them would send a rat running, another five would sneak past.

Through the window I could see that Willie had managed to push my office door shut, ensuring that no dog could deflect him from his twisted mission. He was sitting on his haunches on my desk, flipping the desk lamp's light switch on and off with his paws, sending out his diabolical message to all his little nether-minions. His face was frozen into a maniacal rictus.

"Should have installed those compact fluorescent bulbs," LaVonne said, calmly. "They don't turn on and off quite so fast—it would really throw off his rhythm."

"Yeah. What was I thinking? I should have known this would happen."

Ignoring my sarcasm, she said, "You go take care of your cat. I'll handle things out here."

I started to object, but then I remembered the spooky way she had with animals. She probably didn't need my help. But, just in case she had overestimated her abilities, I ran quickly for the house to hog-tie my seditious cat. As I rounded the corner I saw a pair of packrats near the side of the house, trying to chew through the PVC conduit that housed the heavy wires from the wind tower. A quick glance to the west unveiled three more nibbling on the conduit coming from the solar array, and another half-dozen working on the array's plastic junction

boxes. I pulled off my jacket, swung it around my head, and swatted at them until they all disappeared into the shadows. I unlocked the door and ran for my office.

If Willie was surprised to see me, he didn't let it show. The leer I'd seen through the window was still hard-set on his face, as though cemented in place by Igor, the hunchback taxidermist. However, the second I screamed, "Prepare to die, you good-for-nothing cat!" and lurched for his scrawny neck with outstretched fingers-become-claws, he realized the party was over. Like a furry ball of yellow lightning, he leapt from the desk—scattering hundreds of pages of the novel I'd been working on—zipped out of the office, and ran through the wide-open front door. Though it was obvious—even to me, in my advanced state of rage—that I would never catch him, I took off in hot pursuit.

The ruckus outside had grown several degrees more bizarre since I'd stormed into the house to put a stop to Willie's underground broadcast. As I expected, the big yellow cat was nowhere to be found. But neither had the swarm of packrats returned to bedevil my wiring with their insidious incisors. Looking up, I quickly saw why. The whoosh, whoosh of scores of wings cut through the chilly air, and the night sky was suffused with the darting, gliding silhouettes of a multitude of raptors. It was an aerial feeding frenzy. Dozens of owls and hawks, falcons and kites—even a pair of eagles—had come (or been summoned) from every direction to gorge themselves on the unnatural glut of rat flesh. Each bird hunted according its genetic wont—some spiraled down in low arcs, snatching their rodent victims in singular, smooth sweeps; others dove from the sky like kamikazes, landing squarely on their prey and corralling the hapless rats in their strong, majestic wings before latching tightly onto them and taking off again.

It was all over in just a few minutes. Every packrat that was still breathing had fled to safety amidst the junk and refuse from which it had emerged. I doubted seriously they would return anytime soon.

As the sky cleared and the stars returned to silent prominence, I noticed LaVonne standing serenely in the middle of the yard. Mick, Amy, Newt and Stinky all rested beside her, gazing up into the brilliant firmament. Willie was nowhere to be seen.

"Are you alright?" I asked, jockeying with a pair of dogs for a place beside her.

"Never better," she answered, with a starlit smile, as I put my arm around her and kissed her on the cheek.

"What just happened?" I had to ask.

She wore a distant, dreamy expression, as she said, "Oh, just Nature doing what She does best."

I didn't bother to ask her what part she'd had in the spectacle, or even what her answer meant. It was not the type of answer that begged for qualification; surrounded in a cloak of mystery, it was nonetheless complete. It was just the way LaVonne saw things: to her, the most wondrous pearl was the one left inside the oyster.

"You didn't happen to see if one of the owls carried off a big yellow cat, did you?" I asked, half hoping.

"Oh, no. He ran off over there, somewhere," she said, pointing into a thicket of mountain mahogany.

My first thought was to take the dogs and go flush him out, but I knew he would just run somewhere else. The cat wasn't going to be caught until he was ready to turn himself in. And that wasn't going to happen until he got so hungry the thought of food weighed heavier on his warped little mind than the threat of punishment. But by then, he probably knew, I wouldn't even be mad at him anymore.

I was startled from my thoughts by a whooshing sound overhead. At first I thought it must be a couple of birds conducting a cleanup operation, but I quickly realized it was the propeller blades on the new wind generator reacting to a sudden gust of wind 60 feet up. A moment later we all felt the wind at ground level. It quickly grew in intensity as the generator's blades began to turn ever more furiously. Though the sound was a bit distracting, I quickly realized a good portion of my electricity would now be coming from the spinning turbine atop that tower. It was an exciting concept. I wanted to run into the house and check the meters, to see just how much electricity the wind was producing.

As I turned to go, there came from the bushes a sound quite unlike any that I had ever heard before. It was coming from a cat, that much was certain, though it was not the sound a normal cat would make under any circumstance I could imagine. Definitely not hunger or fear, and probably not even lust, ecstasy or torture. It was a sound as melodious as it was strained; as compelling as it was disturbing. A sound that would've drawn Odysseus and all his men to their doom, had they been possessed of feline faculties.

Wild Willie From Outta Town, it seemed, was not yet through for the night. I looked first to LaVonne, and then to each of the dogs. They were all mesmerized by the operatic mewing issuing from the bushes. Then my eyes dropped to Stinky. Her head waved back and forth like a stoned teenager at a rock concert, and her nose twitched uncontrollably, as though she could actually smell the sound and it was catnip to her ears. Fearing that she was about to be drawn into another of Willie's traps, I reached down to snatch her up.

Too late.

The second I touched her fur she shrieked and took off like a pea through a pea shooter. I called out to her, but to no avail; she was too far under

Willie's malevolent spell. There was nothing we could do but wait, and listen.

It didn't take long for Willie to change his tune. Once Stinky entered the thicket Willie dropped his sonorous charade and the sounds we heard became those of a good old-fashioned cat fight. Apparently Willie needed to take out his frustrations on someone, and Stinky was the only one around he could pick on. I ran toward the bushes, LaVonne and the dogs hot on my heels. I had a big enough vet bill with my horses and calves; I didn't need another for a chewed and clawed cat.

Before any of us could reach the quarrelsome cats, the ear-piercing noises ceased as Stinky launched herself out of the bushes and up the tree beside the wind tower, Willie close on her tail. Great, I thought, it's going to take tranquilizing darts to get them both down. But, to my amazement, things got even worse. Stinky flew out the top of the tree and onto the wind tower. Now she's stuck, I said to myself. No way can a cat climb a steel tower, rungs or no rungs. I was wrong. By the time Willie launched himself from the top of the tree, Stinky was halfway up the remaining 20 feet of the tower, climbing the steel lattice, paw over paw. With the wind blowing and the propeller spinning, I saw no way either cat was going to make it out of this alive. I put my arm around LaVonne and prepared for the worst. All three dogs had their eyes riveted to the top of the tower.

Willie, taking stock of the predicament he'd worked himself into, lost a few degrees of determination. He clung to the side of the tower and watched as Stinky inched her way toward the top.

When she reached the final rung of the lattice, Stinky could climb no more. Above her, a 3-inch steel pipe protruded 5 feet from the top of the tower. On top of that was fastened the turbine and the fever-ishly whirling blades. Willie loomed below, mayhem boiling in his eyes. Wanting no more of Willie, no matter how unsavory the alter-native, Stinky drew down on her haunches and jumped with every ounce of strength her poor tired legs could muster. It was just enough to propel her to the top of the wind generator, where she came to rest, looking very much like a feline aviator in a wingless crop duster with an oversized prop. Together, LaVonne and I clapped our hands and shouted encouragement at Stinky for her valiant maneuver. Amy gave her an empathetic "Woo, woo, woo."

It was more than Willie, in his deranged state of mind, could take. The hesitation he'd shown earlier was now transformed into a rabid lust for revenge. The horrid grin returned to his face and he shot up the tower like a rock from a catapult, not even pausing as he reached the top of the lattice. He hurled himself toward Stinky, but overshot his mark by several inches. Stinky ducked her head at the last instant and Willie, burdened as he was with far more momentum than he'd bargained for, zoomed right past her and into the spinning blades.

I prepared for a screech, a whack-whack, and a nauseating shower of finely-ground cat parts. But there was only an intense, momentary brightening of the light in my office; the same one Willie had used to hail the packrats to his ill-fated counter-revolution. In the last instant Willie had unwittingly given up all his earthly energy to further the very cause he had fought so hard to destroy.

Sadly, I donned a backpack, flipped on the wind brake, and scaled the tower to retrieve my one remaining cat. 🌾

SIX MONTHS LATER...

 so many changes
could it be the way of things?
nothing more, nor less

It was a great time to be alive. Once the blackouts became so wide-spread everyone with even moderate sensibilities knew what was happening, the media was finally forced to admit they'd known the true cause of the blackouts all along. The public outrage was swift and predictable. The powerful airway moguls who had formerly shaped the world by suppressing, inventing, twisting, watering-down and creating the news, were chastised into simply reporting it.

One of the first things the new breed of reporters covered was the press conference during which Gilbert Gigabucks, CEO of Planet Power Corporation, announced to the world that 95 percent of his company's profits would go into research on solar and wind energy, and hydrogen fuel-cell technology. He claimed his company was "firmly committed" to being pollution-free within ten years. Fearing a direct hit to their expansive-yet-dwindling portfolios, many of the larger stockholders cried foul and attempted to have Gilbert replaced, but their protests were quickly muted when another surge of animal-induced sabotage—the biggest one to date—brought worldwide commerce to a screeching halt. Swarms of gnats, in a giant, well-coordinated assault, had worked their way into the mainframes of every Planet Power control station in the country. The computers that once controlled the nation's power grid had been converted by the frolicsome little insects into Pacman play stations. The nationwide blackout lasted for five days. For most of that time all the plant managers across the country, feeling they could deal with the crisis on their own, feverishly played Pacman day and night. Their efforts proved futile. The lockout finally ended when Gilbert Gigabucks enlisted the aid of an army of twelve-year-olds, who swiftly brought the increasingly difficult games to a victorious conclusion. It gave a new, earthier meaning to the term "computer bug."

Following the blackout, the President of the United States, who just a month before had denounced the animal kingdom's concerted effort to shut down the grid as "the single greatest threat to national security this country has ever known," announced that it would now be his administration's "highest priority" to free the country forever from the "horrendous environmental costs" of a coal and oil-powered economy. To show his good faith, the President proudly unveiled a few projects the Pentagon had been working on under the cloak of national defense. The most impressive item, a portable, cooler-sized, solar-powered hydrogen fuel-cell recharger, was years ahead of anything on the drawing boards. It was rumored to have been a gift from aliens, in exchange for a small piece of real estate inside a mountain near Sedona, Arizona, where the extra-terrestrials wished to build a new base of operations.

Personally, I think it originated in Hank's mysterious metal outbuilding.

But who knows? Hank wasn't talking, and I was starting to believe that anything was possible. After all, the Perfect Woman I first met in my dreams now shared my name, and dreamed beside me every night.

Unless, of course, we were awakened by Willie chasing Stinky across the bed after the lights were doused.

Willie? That's right—it seems our grief over his demise was premature.

Mick announced the cat's resurrection in the wee hours of the morning. With an uncustomary paucity of words, the dog said only, "He's back," before disappearing again into the shadows.

The bright flash of light we saw from my office as Willie was consumed by the spinning blades must have been the demon inside of him burning up, at last. As I discovered the next morning, the rest of my wranglesome cat had traveled through the power lines and into the batteries, from which he emerged, howling and scratching at the inside of the battery box cover.

He was tired, confused and rather subdued, but still Willie, all the same. Minus, it seemed, any knowledge of the psychotic episode he'd experienced the night before. But one never knows; with Willie, it could be just be another of his ruses.

All I can say for certain is there was never another instance of packrat sabotage; at least not around here. Of course, stories abound of more unnerving occurrences in other places and, if these harrowing tales are to be believed, I'd say LaVonne and I got off pretty easy.

But that's another story. 🌿

As we were preparing this book to go to press, we spotted this story in the *Rocky Mountain News* Business Section:

SQUIRRELLY OUTAGES
February 5, 2003

Longmont Power & Communications, which serves 35,000 customers, says that more than 90 percent of its significant outages are caused by squirrels. The pesky animals cut the power 393 times in 2002, up from 349 two year earlier. Banding utility poles with slippery, hard plastic apparently didn't help.

It lends credence to my personal Fiction Writer's Maxim: "Never spin a tale you don't want to come true."

Beyond the Grid

## chapter 1

A New Philosophy of Freedom
learning to think beyond the grid

Starting With Nothing

The preceding section was, of course, fantasy. It was contrived to demonstrate the basic principles and components of a renewable  energy system in a fanciful way. This was to keep you from getting bored, shoving the book into a dark cranny, and forgetting about it. To set the record straight, LaVonne has never owned a candy apple red '64 Corvette—hydrogen-powered, or otherwise— though she certainly would like one. Nor did she know as much about renewable energy applications, before we met, as you might have been led to believe. As far as the animals are concerned, Mick's grammar and diction, in truth, leave a lot to be desired. Willie, on the other hand, was presented pretty much as he really was, though, as I recall, he never could get the hang of Morse code.

So much for whimsy, I say. The naked truth of things is never so neat and idealistic—or quite as fun—as the tales we wordsmiths forge on our key-boards to get people's attention. In the fable, for instance, LaVonne and I made the transition from grid power to a fully-evolved solar and wind system in the blink of eye. In reality, we didn't have a finely-honed solar and wind system until long after we sold our ranch, moved into a small mountain cabin, and built a new log home to put it in—a period spanning more than two years.

Once we finally did turn our backs on ranch life for the last time, we quickly realized we were facing a sudden and irrevocable change in lifestyle. The fact that we had just sold a 2,000 square foot house, so we could live in an amenity-free cabin with a footprint smaller than a one-car garage, was the least of it. The real challenge was learning to deal with the sum of all the other changes.

Where before we had a well that pumped water at the rate of 30 gallons per minute, we now had to drive 20 miles to town once a week to fill a 200-gallon tank roped into the back of a pickup. Our toilet was an outhouse, our bathtub a creek. A large cooler had replaced our spacious refrigerator (sorry, no more ice maker). We now used a wood stove for heat, a gasoline camp stove for cooking, and kerosene lanterns for light.

And when we simply *had* to have electricity, I would go solemnly into battle with a war-hardened, 4,000-watt Coleman generator. If I prevailed over the surly beast, then we could run a saw, or a vacuum. If I was bested, then we just had to wait until I could sneak up on the ornery brute in its sleep and yank its cord before it had a chance to suck in a carburetor-full of gas, belch flames, and flood itself.

One would think that such an abrupt "lowering" of living standards would manifest itself as individual stress, or even marital strife, but nothing of the sort occurred. Quite the opposite, in fact; more than anything, LaVonne and I embraced our new life with all the energy and ambition of a couple of kids on an extended camping trip. Having purged our lives of the leaden inertia that crystallizes in the consciousness of anyone in the habit of having an instant remedy for any earthly desire, we quickly came to appreciate every drop of water, every morsel of food, every ray of light in the midst of darkness.

The First Precious Amenities

We were, quite literally, starting over from scratch. Any comforts or conveniences we hoped to enjoy would have to come as a direct result of hard work and mutual cooperation, not from wringing our hands over the nature of our plight.

Living just a notch or two above primitive, we quickly prioritized our desires. (I say "desires" because we actually didn't *need* anything more than we had.) First on our wish list was hot water. And not the kind you heat on the stove in a brass pot for a cup of tea; we wanted hot water flowing copiously from a showerhead, inside a closed shower stall, within a warm room, walled off from the clouds and the wind.

Toward that end, we built a small (6-foot x 16-foot) addition on the back of the cabin; just big enough to hold a tiny shower stall, a used 35-gallon propane hot water heater, a propane refrigerator (which, as it turned out, wouldn't arrive for several months, thanks to Y2K), three water barrels, and everything we needed for a simple solar-electric system.

We pressurized the fresh water system with a small 12-volt RV pump, which was powered with a pair of 12-volt deep-cycle batteries; one to power the pump, while the other was hooked to a small solar module for charging. When the battery in use ran down, we switched them out. (We have since installed a DC to DC converter to run the 12-volt pump from the main 24-volt battery bank, as our stand-alone deep-cycle batteries have at last worn out.)

The electrical system followed the plumbing and gas piping. Using the solar modules and power inverter we'd purchased for the log house—which, at that point, was no more than a muddy hole in the ground—we had more power than we could've ever used in that small cabin, even during a run of cloudy days of biblical proportions.

Considering that, until then, my most meritorious achievement as an electrician was running AC power to a few stock tank heaters for my horses, the complexity of that first, bare-bones PV system seemed daunting. I read every manual, front to back, one, two, three times. I put every component of the system into place, and then slowly, meticulously, hooked-up my wires, using a multimeter to check and double check every step along the way.

I will never forget the day when that first solar electric system came online. I tested every connection ten times over, before I finally flipped the switch to the DC disconnect and turned on the inverter. No sparks, no explosions. Just a steady hum and the soft, green glow of the display. After testing the AC side of the system—to see if my multimeter was as convinced as I was that we actually *were* producing usable power from the sun—I began trying different loads. First a small light, then an electric drill.

Impressed, but still not convinced, I opted for the ultimate test: a voracious 15-amp table saw. It was the single piece of equipment that could make the burly old Coleman generator convulse with fear (of a herniated head gasket, I would imagine). I plugged it in, hit the switch (after a fleeting, jumbled moment of pensive hesitation) and watched with consummate awe as the blade spun quickly and

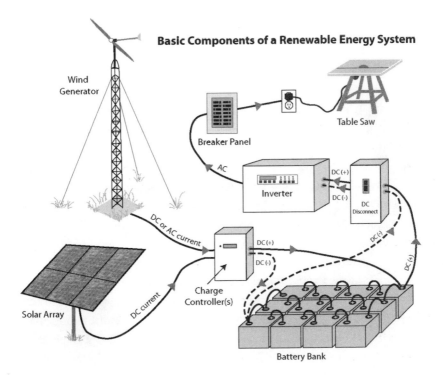

Basic Components of a Renewable Energy System

Wind Generator

Breaker Panel

Table Saw

AC

Inverter

DC (+)

DC (-)

DC Disconnect

DC or AC current

Solar Array

DC current

Charge Controller(s)

DC (+)

DC (-)

DC (-)

DC (+)

Battery Bank

effortlessly into motion. Though my left brain knew the whirring saw blade was merely the logical outcome of applied technology, my right brain insisted I had just witnessed a miracle.

Regardless of which interpretation one chooses to embrace, it was in that instant when the table saw came to life that the idea of sustainable, free power from the sun was transported from the theoretical realm to the practical.

Moving Up

By the time we were far enough along on building the new log house to run wires and install the photovoltaic (PV)/wind system, I thought I knew practically everything there was to know about solar energy. It was a thoroughly absurd notion, of course; kinda like the teenager whose first, distant glimpses of adulthood lead him to overlook life's myriad subtleties and draw the erroneous conclusion that life is a simple subject. (And any grownup who doesn't agree is an idiot.)

The first cracks in my thin, hard shell of ignorance came when our electrician—after not showing up for several weeks—made the grievous mistake of mouthing-off to LaVonne.

He'd have been better off poking a wildcat with a sharp stick.

With the building-boom in town, there was zero chance of finding another electrician who would be willing to drive up into the hills anytime soon. And we couldn't wait. So, (very) reluctantly, I told my wife that, given enough time—and enough books on the subject—I could finish wiring the house, and install the solar and wind systems. And (I nearly choked on this part) do it all to code. I didn't have much idea just what the National Electric Code was, of course, but I was pretty sure I was about to find out.

MICK'S MUSINGS

Watching Rex install the solar and wind systems was instructive. I learned a whole bunch of new words.

On my first inspection after countless hours of work, the inspector wrote me up for eighteen violations. He later called back and admitted he was wrong about two of them, leaving me with only sixteen violations to deal with.

In an attempt to ensure that my next inspection wasn't an encore performance, I must have ended up talking to every high-level electrical inspector in the state, at one time or another. No one remembered my name, but they all knew "the guy with the wind tower," since it was one of the sticking points in our negotiations. (Code requires all sources of electrical power to have a manual disconnect, while design requires a wind generator to be connected to a load at all times. In the end, design won out over code.)

Happily, my wiring passed the next inspection without a hitch; for the first time in two years, we were living in an electrically correct house.

It was only then that my real education began.

Learning the System

Once we moved into the new house, we realized that even after more than doubling our generating and storage capacities we now possessed the means of using energy faster than we could produce it. The well pump was the main culprit, followed by the dishwasher and the hot water circulating pumps for the propane-fired radiant heat boiler. (The usual suspects, namely the refrigerator, range and clothes dryer, were all powered by propane.)

It quickly became clear that a means to monitor our energy usage was necessary—for peace of mind, if nothing else. We checked around to see what was available, then bought a TriMetric Meter from Bogart Engineering. It was a little tricky to install and calibrate, but well worth the trouble. Not only can we use it to see roughly how much wattage each of our

TriMetric Meter
Photo courtesy of Bogart Engineering.

appliances is consuming, it also keeps track of amp hours going into the batteries, versus amp hours going out, providing a digital fuel gauge for the system. (The TriMetric performs a lot of other useful functions, as well. It's worth its weight in gold, though, thankfully, it cost quite a bit less.)

Once the batteries are full, the charge controllers for both the wind and solar systems back off the power delivered to the batteries. It is, therefore, most efficient to use all the energy you can while it's there for the taking. Make hay (wash clothes, run the dishwasher) while the sun shines, as the expression goes. Ideally, we don't want the batteries to become fully charged until the end of the day.

Seeing how the wind plugs into our energy equation has been as fascinating as it has been instructive. While the solar array is the real workhorse of the system, the wind is like a whimsical sprite that always seems to show up, just when we need it most. Many people have remarked that extra solar modules would have been cheaper, delivered watt for delivered watt, than our wind generator and tower. And they're right. But they also completely miss the point, since the wind most often provides power at night and during stretches of cloudy weather, when the solar array is idle, or nearly so. This means that we can get by with less storage capacity than any of our solar-only neighbors, since we're charging our batteries while they're depleting theirs—a fact that is particularly gratifying after three days of cloudy, windy weather.

Daily I feel more respect for our wind and solar system, and its remarkable ability to rejuvenate itself. I have seen our bank of twenty batteries go from 65 percent charge to 100 percent in a single day. And that's while running two computers and a stereo all day long (plus any other appliance LaVonne can throw into the mix).

Consequently, my attitude toward energy usage has become much more relaxed, since I know that we'll gain it all back, in due course. LaVonne, too, has taken notice of my moderated vigilance over the wattage reserves. She hardly ever calls me an Energy Nazi anymore.

It's just a matter of learning to trust Mother Nature.

chapter 2

Grid-Tied or Off-Grid?

should you jump in head first, or just get your feet wet?

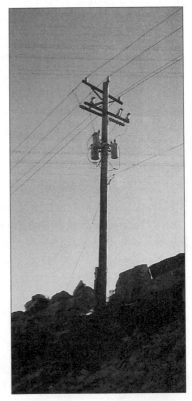

The desire to live off-the-grid is not universal in its urging. For some, it's an extension of a natural passion for self-reliance and simplicity; for others, it's simply the most cost-effective solution to a nagging problem. Most of us fall somewhere in the middle. We know people living within a stone's throw of a utility pole who nonetheless refuse to be hooked into the power grid. We commend them for their tenacious adherence to principle.

For LaVonne and I, it was probably cheaper to go with wind and solar than to tap into the neighbor's power line, 2,000 feet away. But we'll never know, because we never asked the power company what it would cost to run power to our new house. We figured we'd already given them enough money for one lifetime, and suffered through enough blackouts to last through eternity. It was time for something new.

In 1999, we didn't realize the serendipitous nature of our decision. We just thought it was a cool idea. We didn't know that log homes (as our new home was) require far fewer natural resources to build than conventional homes, or that they are warmer in the winter and cooler in the summer. We only knew we wanted to build our own from the ground up, and live in it.

Nor did we realize that, by powering our house with renewable sources of energy, we would, each year, prevent the consumption of over 7,000 pounds of coal on our behalf. This coal, when burned, would release nearly 15,000 pounds of carbon dioxide (CO_2) into the atmosphere; an accumulative tonnage of CO_2 that would, in less than 20 years time, grow to outweigh a cube of solid concrete over 13.5 feet on a side.

We've both come a long way since then; this type of enterprise has a way of changing a person. We've learned a lot about energy efficiency and the utter practicability of homegrown electricity. We've learned new skills, and new ways to think about old problems. We've learned how to conserve when there is little, and how to better use whatever is in abundance. Best of all, we've learned that two people, working alone, can build a beautiful home, power it with the wind and sun, and live in it as though it were a palace.

Off-the-grid living is not for everyone, of course. In most areas of the country electric power is still relatively cheap and readily available, and many people live in homes with energy demands that preclude a wholesale conversion to renewable energy sources. It's hard to shell out hard cash for an expensive renewable energy system when all the quiet and (usually) hassle-free power you'll ever need is right at your doorstep.

MICK'S MUSINGS

Living off-the-grid is nice, but I really miss the toothsome treats the meter reader handed out when I sniffed his crotch and growled.

For those of you who are comfortably hooked into the grid, yet still find yourselves with this book in your hands, a partial conversion to alternative energy sources may be your ticket. But the question still looms: what is your rationale for using renewable energy? If you're worried that cheap, reliable power may soon be neither cheap nor reliable, your concerns are shared by the tens of thousands of Californians that installed some form of solar electric power after the 2000 – 2001 rolling blackouts. I can sympathize. The thought of being without electricity is a powerful motivator, especially in northern climes where electrical power, or the lack of it, can sometimes mean the difference between life or death. Having used my whimsical Coleman generator to alternately heat two homes and pump water for forty horses during a five-day blackout following a spring blizzard, I readily admit to a deep-seated paranoia about not having electricity.

But even if your reasons are purely ideological, you'll still need to decide on which form of grid-tied system you want, though it shouldn't take too much thought. Basically, your choices are two: with or without. Backup batteries, that is. Both of these options will be examined in detail below.

Grid-Tie Systems: Marriages of Convenience

For those of you who do not plan to live way out in the boonies far beyond the nearest power line, but would still like to be on the front lines of the renewable energy revolution, a gird-tie system is the perfect way to get in on the action. Grid-tie systems are designed to run your house with homegrown power from a solar array and/or a wind generator, with power from the local utility company picking up the slack when your

A homeowner in Bend, Oregon recently installed PV panels above his garage and a PV Powered grid-tied inverter. *Photo courtesy of Sunlight Solar Energy.*

demands exceed your generating capacity. And, should your personal generating sources produce more power than your loads require, the extra power can be—in most cases—sold back to the utility, who will then sell it to your neighbor, the one who does not share your foresight and highly-evolved sensitivities.

Grid-tie systems can be configured over a very wide range of wattage and voltages. The system you choose will be largely dictated by your enthusiasm for the venture, and the thickness of your pocketbook. No matter what size system you end up installing, there are (as I mentioned above) two basic types of intertie systems: those with batteries, and those without. If it's important, or essential, that certain electrical loads in your home continue operating when the grid power goes down, then you'll need to have batteries.

Grid-Tie Systems Without Batteries

A grid-tie system without batteries is simplicity, itself. This is because the technology involved is so gratifyingly sophisticated it requires virtually no effort or concern on your part to keep it working. (Did the word "virtually" raise a flag? It always does with me. Just the same, I used it because you still may have to sweep snow from your solar array, or adjust it for the seasonal movement of the sun. Hardly a big deal.) The heart of this system is the inverter. It will take power from your DC sources (solar array or wind generator) and use it to produce an AC output comparable to—if not cleaner than—that supplied by the local power company.

This type of system is far cheaper and easier to maintain than one that incorporates batteries, but it does come with one drawback: when the grid goes down, you go down with it. "But wait a minute!" you say. "What if the grid goes down in the middle of a sunny day? Don't I still get the power from my solar array?" No. Sorry. This is one of the many safety features built into grid-tie inverters. It's to keep the hapless utility worker from getting unwittingly fried by the output of your vast solar array while he's working on the lines. It's a really good idea, once you stop to think about it.

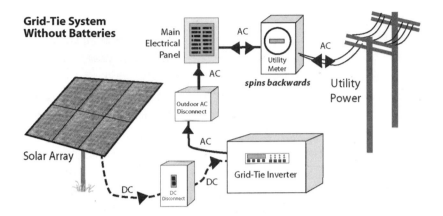

Grid-Tie System Without Batteries

Main Electrical Panel

AC

Utility Meter
spins backwards

AC

AC

Utility Power

AC

Outdoor AC Disconnect

AC

Grid-Tie Inverter

Solar Array

DC

DC Disconnect

DC

Grid Tied Systems With Batteries

Batteries in most grid-tie systems have it pretty easy, compared to those in off-the-grid systems, and the less batteries have to work, the less work they are to maintain. So if the very word "battery" makes you cringe, don't let it. If you really want to make it easy on yourself (and who doesn't?), maintenance-free (sealed) batteries are highly-practical for the first of the two battery backup systems I'll discuss, since the batteries will have very little work to do, as long as the grid is up and running. They can make your system pretty much trouble-free.

Besides the batteries themselves, an intertie system with a battery backup requires the addition of a charge controller between the DC source (PV array or wind generator) and the batteries. This is to condition and regulate the charge before it reaches the batteries. Extra safety components, such as breakers and disconnects, will also be needed, along with an AC sub-panel.

■ Battery-Backup Systems

There are two basic configurations to a battery-based system. In the most popular setup, the battery bank serves as a backup for when grid power is temporarily lost. You may be wondering how this works, since I just told you when the grid goes down, your power goes with it, to keep power out of the lines that utility workers may be handling. Won't battery power go through the lines the same as power from the solar array? No, it won't. The reason it won't is as ingenious as it is simple: when the grid is down your inverter will only provide power to critical circuits (such as the furnace, refrigerator, and a few lights) which are separate from the rest of the system. These circuits will be located in a separate sub-panel that the inverter keeps isolated from the outside utility lines. So, when the utility power is out, you won't have all the amenities you are used to, but at least you won't freeze in the dark while nibbling on rotting food.

During peak generating times, the inverter will first power the house loads with energy from your solar array and wind generator, and then charge the batteries with whatever is in excess. If the batteries are fully charged and the loads are taking less current that the charging sources provide, the excess will be sold back to the power company. Slick.

A website that lists state-by-state information about incentives and net-metering: **www.dsireusa.org**

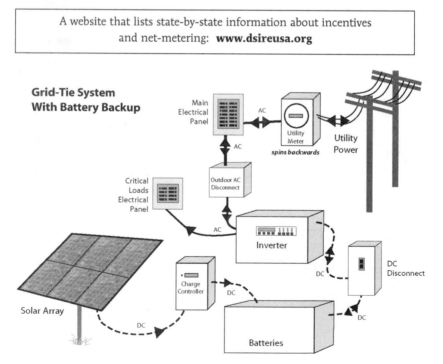

Grid-Tie System With Battery Backup

Main Electrical Panel

AC

Utility Meter

spins backwards

Utility Power

AC

Critical Loads Electrical Panel

Outdoor AC Disconnect

AC

Inverter

DC

DC Disconnect

Charge Controller

DC

DC

Solar Array

DC

Batteries

◼ Grid Backup Systems (Grid Parallel)

A variation of this system is one in which the home (all or part of it), is set up as an off-the-grid system that would normally have a generator backup. The difference in this system is that grid power replaces the fossil-fuel-fired generator. Grid power is not used until the inverter determines that the batteries have reached a critically low voltage threshold. When that threshold is reached, the grid power will kick in and charge the batteries back up to a pre-programmed level before disconnecting.

This latter system generally requires more batteries than the former, and the batteries will be worked much harder, since the system they are a part of is off-the-grid 90-plus percent of the time. For this reason you should plan on using wet-cell, lead-acid batteries (the kind you add water to) since hard work and lousy pay is what they're designed for.

Nor will you be able to sell power back to the utility since, for obvious safety reasons, the system is configured so that power can only flow from the utility, not back into it.

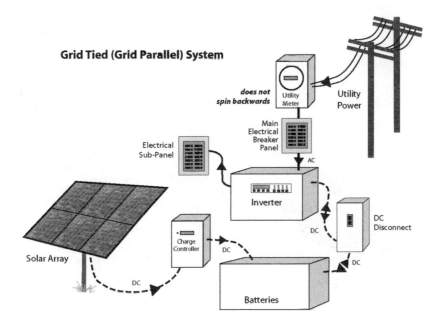

Grid Tied (Grid Parallel) System

Doug's Net Metering Setup

The 48 Uni-Solar peel-and-stick panels on Doug Pratt's garage are nearly invisible, but the 20 kWh per day of electricity (on average) that it provides is quite noticeable, as is evidenced by the zero electric bills from his utility company. In Doug's own words:

"We installed a battery-based system because lengthy power outages are not uncommon in our rural area, and with the nice California rebate we could afford to spend a bit more for the security. In my power room foreground you see the sealed batteries with a wooden cover so nothing can get dropped on them. On the wall, there's pair of OutBack 3,600-watt inverters with a DC box and a pair of the superb MX60 charge controls on the near side, and an AC box on the far side for all the wiring and safety gear. Further down the wall there's a pair of standard circuit breaker boxes.

One is for utility power, the other for inverter power. Note they're connected by a gutter box so individual circuits could be easily moved if we change our mind about what wants backup power.

"In our present home we use about 8 to 10 kWh per day. When we build our new energy-efficient, passive-solar home, we'll use a ground-coupled heat pump for heating and cooling. Even at regular electric rates these pumps are the least expensive way to heat and cool a house, but with all of my extra free electric power, it'll be no problem to run a pump! Electricity, heat, and cooling with no impact on my children...I can live with that!"

Doug Pratt is co-author of Got Sun? Go Solar, *and former technical editor of* Real Good's *Solar Living Sourcebook. Photos by Doug Pratt.*

Choices, Choices

Which way should you go? Systems without batteries, while vulnerable to blackouts, are 3 to 5 percent more efficient that battery-based systems, since there is no transition of electrical energy into chemical energy, and back again. Generally, these systems are installed in places where the grid is fairly reliable, and rebates, tax credits, and other similar inducements are there for the taking. Battery-less systems are often viewed as long-term investments. By the time your system is paid off, you've enhanced the value of your home and reduced your electric bill to little or nothing, while creating an iron-clad hedge against rising utility prices.

Another important factor is the price your local power company will pay you for your excess power. Some states have passed laws requiring the utility companies to buy back your power for the same price you pay for it. This is called "net metering." If you're lucky enough to live in an area where net metering is available, then it makes sense to install a system that sells excess power to the utility during the day, and buys it back at night. If you should happen to live in a state like California, where Time-Of-Use (TOU) billing is in effect, you will be able to sell your excess power to the utility for as much as 32 cents per kWh during the day, and buy it back for as little as 9 cents at night. It's like playing the stock market every day, knowing you'll be getting a nearly 4 to 1 return on your investment.

In other areas, however, the deal is not so sweet. Wherever the utility pays only their "avoided costs," you can figure, roughly, that every two watts you produce will buy one of theirs. In that case, you're better off creating, using and storing as much of your own power as you can, because its worth twice as much to you as it is to the utility. Grid-parallel systems are popular in such places, since buy-high-sell-low is hardly a sound investment philosophy.

MICK'S MUSINGS

A sight I'd like to see: The Cat accidentally intertying himself to The Grid.

While many states now allow net metering, some still do not. Before designing a system, check with your local utility to see what their policy is. You should also check state and local electrical codes to see how easy (or difficult) it's going to be to tie into the grid. Do your homework, in other words. That way you'll know which way you want to play the game.

For more infomation about grid-tied systems and your options, I suggest reading *Got Sun? Go Solar,* a book by me and Doug Pratt.

Comparison of Grid-Tie & Grid Parallel Systems

Grid-Tie *without* Batteries
- Requires fewer components than a battery-based system
- Is more efficient than a battery-based system
- Can sell excess power back to the utility company
- You'll be left in the dark when the grid goes down

Grid-Tie *with* Batteries
- Can use maintenance-free batteries
- Can sell excess power back to the utility company
- Essential loads in your home will still operate when the grid goes down

Grid Parallel With Batteries
- Batteries used extensively
- Grid power used to only recharge the batteries (instead of a generator)
- All loads isolated from the grid continue to operate when the grid goes down (as long as the batteries hold out)
- Cannot sell excess power back to the utility company

A New York family installed 24 BP solar modules on their home and a pair of Sunny Boy inverters for their grid-tied system. They also switched to high-efficiency appliances and compact fluorescent light bulbs to further cut their energy usage and become more self-sufficient. *Photo courtesy of the Andersens.*

CU 2005 Decathlon House

When is a house more than a home? How about when it comes equipped to provide all of its own electricity and heat, and then wins the 2005 Solar Decathlon competition. The University of Colorado 2005 Decathlon House did just that. It was shipped to Washington D.C. in September, and ended up taking top honors in a DOE-sponsored 10-part competition with homes from 17 other universities from the U.S., Canada, Puerto Rico and Spain.

What is a Decathlon House? The concept is simple enough. The rules require each of the competing universities to design and build a house of 500 to 800 square feet. It has to look nice and be energy efficient, of course. But it also has to be sustainable—and then some. This means the active and passive energy systems must capture and exploit enough sunlight to supply all the home's electrical and hot water needs. That includes the power to keep the house in a comfortable temperature and humidity range in all kinds of weather, and also to run lights, a refrigerator/freezer, TV, computer, washer/dryer, electric range...the usual stable of appliances found in any grid-tied home, in other words. Oh, and it also has to produce enough extra solar electricity to operate a street-legal electric vehicle.

That'll get you into the competition, but to win it, you have to do it all with a certain sense of flair. And that's where the CU team, headed by engineering grad student Jeff Lyng, excelled. Christening their creation the BioS(h)IP, the CU team developed their very own **Bio**-based **S**tructural

Insulated **P**anel System (hence the acronym) from waste paper and soy insulation.

Left: Jeff Lyng and Seth Kassels install a PV module on the roof. Below: The home under construction shows the PV array covering the roof, and Jeff gives Rex a look at some of the batteries located at one end of the home. *Photos courtesy of CU-Boulder.*

Demonstrating an R-value of R-7 per inch, the 6½" walls have a total insulation value in excess of R-36. High performance windows range from R-8 to R-14.7, and the clerestory aerogel-filled panels allow plenty of natural lighting while providing an insulation value of R-14.

To conserve energy and natural resources, the CU house was constructed using a creative medley of agricultural and forestry byproducts, including: soy, wheat, corn, kenaf (an African hibiscus), jute, hemp, bamboo, flax and waste paper. The preponderance of food products prompted the CU team to refer to their creation as "a house you can sink your teeth into."

But all the clever design and passive solar features in the world don't do a lot of good without the active solar systems that bring the house to life. To address that issue, the students installed a total of 6.8 kW of Sunpower 200-watt panels, a pair of OutBack inverters, a trio of OutBack MX60 charge controllers and 40 (yes, forty) Deka L-16 batteries to power the house and charge up the GEM E4 electric vehicle.

Supplying the hot water needed for domestic use and for the hydronic in-floor heating system, the CU team installed an array of 80 Mazdon evacuated-tube solar collectors (see photo on page 231), manufactured by Thermomax in Northern Ireland. A 200-gallon storage tank holds the solar-heated water, while a 40-gallon electric water heater serves as backup when the sun is being uncooperative.

All told, it's a house any conservation-minded family would love to call home.

Care to learn more about the CU Decathlon house? Visit: *solar.colorado.edu/design/green.shtml*

For an overview of the entire Decathlon competition, go to: *www.eere.energy.gov/solar_decathlon/*

The Solar Decathlon winning home by CU-Boulder, and the vehicle it powered.

chapter 3

Beginning Considerations
building smart to keep the power bill low

Passive Solar Design for Efficiency and Comfort

■ Natural Daylighting with Windows

If you are building a new home, you'll have plenty of options for optimizing passive solar energy long before you hook up your active solar array. LaVonne and I looked at a lot of houses before designing ours. We were thoroughly amused at the evolution the American home has undergone in the last 50 or 60 years, especially log homes. Many of the older log homes we saw back in the woods were little more than fortresses with tight, economic floor plans, low-pitched roofs, and conspicuously few windows, all barely big enough to aim a rifle through. Many seasoned homes in town were not much better. They were built back in the days when windows were considered to be a necessary—but barely tolerable—source of heat loss. These old homes were well designed to conserve any heat produced within the walls—the dearth of insulation and multi-glazed windows notwithstanding—but woefully deficient at allowing in heat from the outside.

WILLIE'S WARPED WITTICISMS

Soft couches and south-facing windows have made cats nature's most efficient solar collectors.

The advent of efficient double- and triple-glazed windows has changed that myopic view of home design. Realizing the potential of free solar heat in the winter, most homes today—outside of those built in cheesy developments—take ample advantage of the sun's gifts. Prow-like projections, with giant windows set between massive frames, adorn the southern faces of many modern log homes.

This is all well and good, but like everything else, it's possible to get carried away. An overabundance of glass will certainly keep a house warm on sunny winter days, but it will also allow excess heat leakage at night, or when the weather turns cloudy. If you are building a new home, you should strive

Our south-side dormer is filled with windows for excellent passive solar (the center window opens for ventilation). The large eaves keep out the intense summer sun.

for a balance. Design your house to allow ample sunshine on the south side, with as few doors and windows as possible on the north. Large windows, glass doors, and dormer (or gable end) windows that follow the contour of the roof all allow plenty of sunshine while enhancing the home's appearance.

Are there any numbers to buttress the intuitive, though somewhat nebulous, concept of window size and placement? In *The Solar House: Passive Heating and Cooling*, author Dan Chiras suggests the surface area of south-facing windows should be 7 to 12 percent of the heated floor space, leaning toward the higher number in more northern climes. He goes on to say that north- and east facing windows should be limited to 4 percent, and west-facing windows should equal no more than 2 percent of the heated floor area. These ratios will maximize winter heating and summer cooling.

If winter is a long, cold ordeal where you are planning to build, you might want to consider using triple-glazed, or super-insulated windows. They allow as much light to enter as conventional double-glazed windows, but will retain appreciably more heat when the sun sets. Insulated window coverings are also extremely helpful for night-time heat retention.

■ Orientation of the House

Unless you plan to build a dome house, or some other design that deviates from conventional building practices, your house will have a long axis and a short one. For the purposes of utilizing passive solar energy—and to make things easier on yourself, once you get around to installing active solar PV modules and collectors—you will want to run the long axis of the house east and west to maximize the amount of solar radiation that enters the house in the winter months.

If you are a few degrees off it really won't diminish the efficacy of passive or active solar systems. So if you know the magnetic declination for your area (since magnetic north and true north differ from one another by varying degrees in different places) you can use a compass to get a pretty close approximation of the cardinal points. On the other hand, if you are truly persnickety—I know I am—about the alignment of things, you can determine the true north-south axis with a couple of T-posts and a clear night sky (see 'Where is North?', page 122).

With the long side of your house facing south, the area exposed to direct solar radiation in winter will be greatly increased. There are a couple of things you can do to make the best of this free energy. An open floor plan will allow solar radiation greater access to the deep recesses of your house. And by the judicious placement of mass within the sun's path, heat can be stored up during the day and released back into the room at night. Stone, tile and concrete floors are particularly efficient in this regard (and won't fade in sunlight like wood floors and carpeting). In addition to flooring, these materials can be used in planters, islands, and free-standing fireplaces—all creative projects you wanted an excuse to build, anyway.

■ Big Eaves

Big eaves are a practical addition to a new home, since they keep direct sunlight from hitting the windows in the warmer months. Then, in the winter when the sun drops low on the horizon, an abundance of solar radiation brings welcome warmth on cold, sunny days. Another bonus of big eaves: they will protect the sides of your house from rain and snow.

■ Cooling

What about cooling? Standard air conditioning is far too power-hungry for a home powered solely by solar- and wind-generated electricity, and counterproductive for grid-tied homes. Evaporative (swamp) coolers use much less energy than air conditioners, and in dry climates, they add much-needed moisture to dry air.

The same windows that heat your home in the winter can help cool it in the summer (providing, of course, the house has the above-mentioned big eaves). The key is to provide adequate cross-ventilation, preferably of a type that allows air to enter near the floor, and exit through the roof. Double-hung windows, or windows designed with built-in vents at the bottom are perfect for letting air into the house. To provide an exit for hot, rising air, skylights—the ones that open—can't be beat; they are perfect for maximizing airflow through the house. Skylights also help to bring additional, natural light into a house, something that is often needed in lofts with few windows. Just don't install too many of them, or you'll defeat their purpose. Unlike windows under big eaves, skylights are great for letting in summer sunlight and letting out hard-won winter heat. (Fortunately, this problem can be minimized by purchasing skylights with blinds.) Many homeowners prefer north-facing skylights which prevent the scorching summer sun-rays from letting in too much heat.

Skylights are excellent for natural daylighting and ventilation. *Photo courtesy of Velux.*

For those living off-the-grid, ample natural light is a must. LaVonne had me put a small skylight in the loft walk-in closet, and she very rarely needs to turn on the light as she fishes around, looking for just the right ensemble (I guess that's what she's doing, anyway).

Non-Living Spaces

Every house has a mechanical room, where the furnace, or boiler, and the water heater are located. If you are building a new home back in the woods and plan to pump water from a well, you may also need to allow space for a cistern, or a pressure tank. Basements are great places to hide all the stuff you don't want to look at.

■ Electrical Room

In the house where you are now living, there is a single conduit with two heavy wires running into your main electrical panel. This conduit runs from the utility grid to the outside of the house, through a wall, to the panel. It's hidden in the wall and takes up no living space. Things are a little different in a home powered by renewable energy.

In grid-tied systems without batteries, the inverter and other components are usually mounted to an outside wall and thus take up no interior space.

If you are planning a grid-tied or off-grid system *with* batteries, your house will also need to have an electrical room, or at least some out-of-the-way place big enough to contain all the components of the biggest solar/wind system you will ever conceivably build. This may include wall space for the inverter(s), DC disconnect(s), charge controller(s), a 120-240 volt AC transformer, and floor space for your battery bank.

To give you a rough idea of how much space everything takes, our electrical room is 9-feet by 4-feet, with a 7-foot, 4-inch ceiling. Our twenty golf cart style batteries fit nicely in a space 48 inches by 40 inches. By squeezing components closer together, we probably could have gotten by with a little less space, but not much. If you can, leave yourself plenty of extra room. In particular, allow yourself space to expand the size of your battery bank, should you want to add more batteries later on (as we did).

Most wind and solar suppliers/consultants will be more than happy to help you design your system, and their catalogs give precise dimensions of every component you'll need in your electrical room, including the batteries. So, once you familiarize yourself with the essentials of a photovoltaic solar and wind system and understand what's what, it's a simple matter to draw a basic diagram detailing where everything will go.

A few other things you'll need to keep in mind:

- Your **battery box cannot be located directly beneath any serviceable component**, such as an inverter (for servicing and safety reasons, according to the National Electric Code). And neither can anything else. It's really too bad, because that's the logical place to put the batteries, since you will want the large cables connecting the batteries to the inverter to be as short as possible. One clever way around this problem is to build the battery box on the *other* side of a frame wall, just *behind* the inverter, and running the heavy cables through the wall. Conversely, locating the battery box just to the side of the inverter will make for a short cable run.

- **Lead-acid batteries**—the ones most used in off-grid and grid-parallel systems—need to be in a sealed box, **vented to the outside.** It only takes a 1-inch PVC vent pipe, but the closer you can locate the batteries to an outside wall, the better.

- **Inverters hum**. Just *how* much they hum depends on how hard they're working. To keep your sanity, put the inverter in a

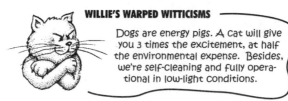

WILLIE'S WARPED WITTICISMS

Dogs are energy pigs. A cat will give you 3 times the excitement, at half the environmental expense. Besides, we're self-cleaning and fully operational in low-light conditions.

room that can be sealed off from the rest of the house in general, and your bedroom, in particular.

- For the sake of efficiency, it's best to locate all of your solar and wind components in the **same room as the main electrical panel**. That way everything is in one handy place.
- All serviceable components (such as inverters) need to have at least **36 inches of space in front of them**. This is to allow plenty of room for ex-defensive linemen to work on them.

Solar and Wind Power Considerations

■ Southern Exposure

Our house is situated on a saddle, 500 feet above a creek, and surrounded on all sides by distant mountains. The sun rises here about an hour later than it does on the plains to the east, and sets an hour earlier. It's about as good a spot as you can hope for in this area. Since the solar wattage available from the early and late-day sun is only a small fraction of what's available at midday, we lose practically nothing to the surrounding hills. At 40 degrees north latitude we have all the sunshine we need to power our house, even in late

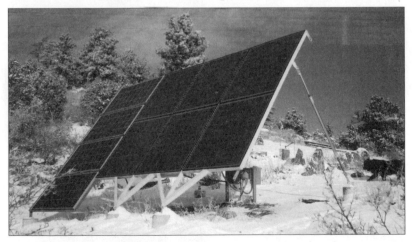

Our ground-mounted solar array is easy to clean, and can be adjusted for the angle of the sun. The winter setting shown here is at 61 degrees.

December when the days are so short that lunch is the only meal we can enjoy in daylight.

Since your solar array will be a major investment, you'll want as much direct sunlight as you can get. If you plan on building in a valley, be absolutely sure your winter sun isn't blocked by the hillside to the south. This can be deceptive. In July, when the sun burns high in the sky like a blast furnace ten feet overhead, it's hard to recall just how far south old sol retreats in winter. Several years ago I rented a little place in a deep valley one canyon north of where LaVonne and I now live. It was bright and sunny in August when I moved in, but as fall approached I watched with dread as the sun crept ever south. Then, early in November, it just disappeared and didn't offer up a single yellow ray until the following February. The sun's arc was nearly identical to the topography of the mountain to the south; one day the sun was there and the next day it wasn't. Fortunately, I didn't own the place and it was wired into the grid, so I could at least see what I was doing as I shoved copious quantities of cordwood into the stove to keep from freezing to death during the three-month shadow.

■ Locating the Array

Plan where you're going to put your solar array. The roof of the house is fine, but modules are much easier to clean, adjust for seasonal changes in sun angle, and sweep the snow from if they are mounted on the ground, or on the side of a deck. Either way, you won't want your array much more than 100 feet from your battery bank, because in low-voltage DC systems the size of wire required to carry the current without substantial line loss increases greatly with distance, and heavy wire can get pricey. It's not much fun to work with, either.

Do you want to know for certain how surrounding trees and hills will affect your array? The **Solar Pathfinder** shows where the shadows will be any time of the year.

■ Thinking About Wind

I'll have a lot more to say about wind in later chapters, but it might be helpful to touch on a couple of points here, just to get you acquainted with the subject.

The wind, of course, blows where it will, and you will either have enough wind to justify the expense of a wind turbine and a tower, or you won't. Some states are windier than others, and every point within each state is different from every other. Secluded valleys are generally not good places for wind generators; mountaintops are great. You could get a ton of expensive equipment and monitor the wind at your site over the course of a year to determine if it blows enough to justify the purchase of a wind generator, or you could follow my simple rule of thumb:

> If the wind blows hard enough and often enough to annoy you, you can probably make good use of a wind generator.

Once you determine—by whatever means—that you do have enough wind to make a wind-generating setup viable, you'll need to start thinking about where to put your tower. This is a little trickier than finding a spot for your solar array, since it will need to be a safe distance from the house (if it's a big system, which is to say, over 500 watts), but not so far away that you need to take out a loan to buy the wire leading to it.

Furthermore, it will have to be high enough off the ground that nearby buildings and trees do not block or dampen the wind from any direction. Ideally, the generator should be mounted at least 20 feet higher than the highest point within a lateral radius of 300 feet. This isn't always possible, but it's a starting point.

Backup Generators for Off-Grid Homes

Unless you've got the unassailable PV/wind/micro-hydro system all the rest of us can only dream of, there will be times when you'll need more energy than your system is putting out. In all likelihood, this energy will be supplied by a fossil-fuel fired generator.

Backup generators are powered by one of three fuels: gasoline, propane, or diesel. Generally, propane and diesel generators are stationary (diesel because of their size, propane due to their necessary proximity to the fuel source), while gasoline generators can be moved from place to place. Many gasoline generators can be converted to propane (and back again), while diesel generators just run on diesel, or one of the bio-derived equivalent fuels available in some places.

The type of generator you choose will be determined by how much you intend to use it, and what you plan to use it for. If you only run it once in while, or like to move it from place to place (to do more than just charge your batteries or pump water from your well) you'd better stick with gasoline. If you can dedicate the generator to the status of backup-power-provider, then propane or diesel will be your best bet.

Buying a bigger generator than you need is wasteful and expensive. Size your generator to run your biggest conceivable load, such as your well pump, and be certain that you can pull that much amperage out of one circuit, because the biggest generator in the world won't do you any good if its power is dispersed between several small circuits.

Altitude matters. All generators are rated at sea level, where there's lots of air. As you leave the beach and move into the hills, the power will drop off at a rate of 3 – 4 percent per 1,000 feet of altitude. It really begins to add up around 6,000 feet.

Unless you're hard of hearing—or have a desire to be—buy a good, quiet generator. Honda generators are especially merciful in this regard. In any event, a quiet generator will be more expensive, more reliable, and longer-lived than a noisy one.

We use a Honda 6,500-watt generator for backup.

While almost any generator with an electric starter can be rewired for remote start (via your inverter, for instance), if you think you will want this feature, it will be cheaper and easier to buy a generator already wired for that purpose. Propane is your best choice for remote starting, since most gasoline generators have a manual choke and do not start easily if it's not engaged, and diesel generators are just plain hard to start in the winter, even with winterized fuel.

And finally, make sure you have a runtime hour-meter. If one doesn't come with your generator you can buy an add-on meter fairly cheaply. They are easy to install and will let you know when it's time to change the oil. And you will change the oil, won't you?

MICK'S MUSINGS

If humans had noses and ears like dogs, their gas generators would spend a lot more time collecting dust in their garages.

Evaluating Your Electrical Demands

To determine what size, or kind, of system to install, it will be helpful to calculate just how much electricity you currently use. This is quite simple—just divide the kilowatt hours from your last electric bill by the number of days in the billing cycle. Unless you are quite conservative, or have a lot of watt-gobbling tools and appliances, it should be in the range of 15 to 30 kilowatt hours per day.

That was easy.

Now comes the hard part. Considering that the modules for a solar array will cost you somewhere in the neighborhood of $3.50 – $5.00 per watt (and each of those watts will give you, on average, .004 to .006 kilowatt hours per day, which is to say that a 1,000-watt array will yield 4 to 6 kilowatt hours on a good day), what are you willing to spend and/or do without to get in on the renewable energy revolution?

Below is a non-exhaustive list of things to avoid (like the plague, for those of you planning to live off the grid) if you want your new venture to produce satisfying results:

- Air conditioning
- Electric clothes dryer
- Electric range/oven *(continued on next page)*

- Electric water heater
- Electric heating (of any kind)
- Older electric refrigerator and freezer *(prior to 1999/2000)*
- Incandescent light bulbs

I'm sure for many of you the thought of getting by without air conditioning is tantamount to doing without food and water. If you are building a new house, you can incorporate design features that promote passive cooling (as mentioned earlier in this chapter) but in certain areas of the country (Houston in July comes to mind) even the best designs will offer little relief. If that's your situation, you should forget about living completely off-the-grid (unless you've got a key to Fort Knox). Everything else on the list, however, can be substituted with an energy-friendly counterpart. You can just as easily use:

- Compact fluorescent light bulbs
- Gas clothes dryer
- Gas range/oven
- Gas or solar water heater
- Gas furnace/boiler
- Wood stove/boiler or masonry stove
- New, efficient, electric refrigerator and freezer

Compact fluorescent light bulbs come in a wide selection of wattages and sizes.

How much will these things lower your energy consumption? Plenty. If you trade out all the appliances on the first list (air conditioning, notwithstanding) with those on the second, you should easily cut your current energy bill by more than half (see the appendix for a list of appliances, and each one's energy usage, as well as a worksheet to calculate your energy usage). If you are serious about renewable energy, I suggest you invest in a *Watts Up?* or similar meter to measure the energy consumption of the appliances you now have, and then compare them to the appliances listed on the Energy Star website (*www.energystar.gov*).

Okay; so (on paper, at least) you've herded all the energy pigs out of your house and replaced them with efficient, state-of-the-art appliances. At last you've got your hypothetical energy usage down to a manageable sum. Maybe, you think, this renewable energy thing will really work. So what else can you do to maximize your investment?

The next best thing you can do is to give away all of your plug-into-the-wall electric clocks and replace them with battery-operated models. They're 100 times more efficient. A plug-in clock will use up to 26 kWh of electricity per year while battery-operated clocks will run for two years on a single AA battery. Go figure. After you change out your clocks, put everything that can draw a ghost load (a small load drawn by an appliance, even when the power is turned off) into a power strip that you can turn off whenever you're not using whatever is plugged into it. This includes TVs, VCRs, DVD players, cable boxes, and anything that has a little black AC to DC converter (such as laptop computers, chargers for cell phones and batteries, etc.). Even a load as small as one watt can add up over time: one watt drawn continuously for 24 hours is equal to running a 1,000-watt hair dryer or microwave oven for 1½ minutes each day.

You can be as militant or as charitable about this as you want, of course. We drew the line with the clocks on the gas range and microwave oven; hence, when I stumble down the stairs in the dark to throw another log in the wood stove, I always know what time it is.

This will give you a good start. I've included other pointers throughout the book, in the appropriate places. Once you invest in a renewable energy system and begin living within its framework (or constraints, depending on your point of view), you will forever amaze yourself with the little tricks you manage to come up with to save a watt or two.

A Watts-Up? meter is a simple way to measure the electrical usage of appliances (a toaster readout of 781 watts, or .781 kWh, is shown above).

---

one watt delivered for one hour = **one watt-hour**
1,000 watt-hours = one **kWh**
amps x volts = watts   *Example: 2 amps x 120 volts = 240 watts*

Personal Power Companies

Homeowners	Lane, Sue, Haley & Finn
# of Occupants	2 adults and 2 children
Location	Masonville, Colorado
Home Size	1,040 square feet plus separate ceramic studio
Home Heating	Lopi wood stove & backup propane wall heater
Water Heating	Propane on-demand
Grid-Tied?	No
PV Array	496 watts ground-mounted; tilt for the seasons
Charge Controller(s)	Trace C40
Inverter(s)	Trace (Xantrex) DR2412 modified sine wave
Batteries	12 deep-cycle, lead-acid T-105 batteries 1350 amp hours @ 12-volts
Wind Turbine /Tower	None
Solar Hot Water	None
Backup Generator	Coleman PowerMate 5000
Comments	Of all the off-grid people we've ever met, Lane and Sue have flourished the longest (since 1991) with the least. Their two children, ages 12 and 5, have never lived in a home with grid power and have no desire to. Besides the cabin's few electrical appliances—laptop computer, TV/VCR combo, small jet pump to pressurize domestic water, and a few lights (the fridge and range run on propane)—Lane also uses their thrifty solar system to run a few tools and to power his large wood-heated ceramic studio.

## chapter 4

Sizing the System

a short, no-tooth-pulling course on practical electricity

Beyond Ohm's Law

The more you delve into the electrical particulars of renewable energy, the more you will hear the term "Ohm's Law" bandied about. This is curious, because electricians—not to be confused with electrical engineers, who actually do use Ohm's Law—hardly ever have occasion to refer to George Simon Ohm's most notable accomplishment; at least not in its basic form.

Ohm's Law (*resistance* = *volts* divided by *amperes*) is very handy for determining volts, amperes, and ohms (units of electrical resistance) when two of the three variables are known. If, for instance, you want to know the amperage of a circuit, you can measure the change in voltage across a conductor of known resistance. Likewise, you can use it to determine resistance and voltage.

But you won't. You are not concerned with the "R" part of the equation. That's the reason you have "wire size/line loss tables" in every book ever written on the subject (including this one). Thanks to Ohm's Law, and a zillion or so formulas based on it, the phenomenon of resistance has been

MICK'S MUSINGS

A Dog's Power Formula: The power of a dog is directly proportional to the square of the number of cats nearby to terrify.

standardized in the electrical industry to the point that it is there without you even knowing it.

Let's put it in human terms: You want to brew a few cups of coffee in the morning. If the people who designed and built the coffee maker were ignorant of Ohm's Law, it would be a real hit-or-miss affair. But they knew all about it, so you don't have to. All you really want to know is how many minutes of direct sunlight it will take to fill your cup with aromatic black liquid. In other words, how are the volts (V) and the amps (I) required by the coffee maker related to the power produced by your solar and wind system? Resistance doesn't do you any good here—you're interested in watts (P). That is, after all, the units used to rate wind generators and solar modules. Fortunately for those us who enjoy mental math, the formula is every bit as easy as Ohm's Law: **P = VI, or: watts (P) = volts (V) x amps (I).**

The above mentioned coffee maker runs on 120 volts, and draws 7 amps, so it will take 120 x 7 = 840 watts to run it. That's simple enough. So how much energy did it use? Well, if it takes 6 minutes to make coffee (and you don't leave the warming plate on after it's done brewing), then that's 1/10th of an hour (.10 hours, for those of you with calculators). So, taking 840 times .10, you require 84 watt hours to scald the sleep out of your brain in the morning. How much energy will it use over the course of a year? If you divide 84 by 1,000, you will see that your coffee maker requires .084 kilo-watt hours per day. Multiply that by 365 (days in a year) and you arrive at 30.66 kWh/year. This is about four sunny, summer days' output on a 1,200-watt PV system, meaning that four days of each year are dedicated to making your coffee each morning. This is about equal to watching TV for 1½ hours each day, so why not just turn off the TV, grab a good book and enjoy your coffee?

There are two other expressions of this equation: I = P÷V is the most helpful, since you can use it to determine wire size from the inverter or charge controller to the solar array or wind generator (among many other places).

On the other hand, since voltage is almost always a known variable, the final expression, $V = P \div I$, will probably be of little use to you, or anyone else. You may find it ironic, then, that voltage will be the one immutable factor in the final configuration of your system.

The Appendix discusses in detail the formula for calculating exact line loss, plus examples on how to use the Wire Size/Line Loss tables.

Power Formula
watts = volts x amps
amps = watts ÷ volts

12, 24, 48 Volts, Or More?

Most residential solar modules are rated for 12 volts. More than likely, the batteries you buy will be 6 or 12 volts each. Either of these components can easily be wired in series to increase the system voltage. If you buy a good charge controller, you can change the voltage from 12 to 24, to 48 or more volts, by changing a single jumper on the circuit board, or pushing a few buttons on the display.

After that, you run into a low (or high) voltage wall. The wind generator you buy may or may not be adjustable to different voltages. And here's the kicker: the inverter—one of the most expensive components in your system—will definitely **not** be adjustable.

It's decision time.

By running the formulas in the Appendix (or taking the easy way out, and referring to the line loss tables) you will see that by doubling the system voltage (and thereby halving the amperage) you can drive the same wattage four times the distance through the same wire! It would seem, then, that a high system voltage would be the answer. And it is; but how high?

If you are looking for a catch, it isn't the price of the components. There is no difference in price between a 12-volt inverter and a 48-volt inverter. The same holds true for everything else. If you want to include a few 12-volt DC circuits in your house, you'll have to buy a step-down converter to pull the lower voltage out of a higher-voltage system, but the money you save on wire will pay for that minor item several times over.

The only real downside in systems over 12 volts is in the multiples of solar modules and batteries you will have to buy (initially, and if you need to add on later). In 24-volt systems, the solar modules are wired in pairs, and the 6-volt batteries are wired four in a series. Double these numbers for a 48-volt system. So, if you have 12 solar modules charging a bank of 16 batteries in a 48-volt system, and you want to increase the system size, at a bare minimum you'll have to up the modules to 16, and the batteries to 24. It can get expensive, especially if you are using the big 390-amp hour, L-16 batteries. (It is possible, however, to run a 24-volt system with a 48 or 60-volt array. *See Chapter 8, Charge Controllers.*)

A 24-volt setup is a good compromise for a medium-sized system, if you don't have the physical space or the financial resources to enlarge your system in such large chunks. You can add onto it without breaking the bank, and still get 4 times the current-carrying capacity from your wires, over a 12-volt system.

Is it possible to design a battery-based system over 48 volts? Sure, but it's hardly a practical solution for a private residence. Besides the fact that components are hard to come by, such systems are expensive, inflexible, and far more difficult to build within the constraints of the National Electric Code (NEC). On the other hand, many direct-tie systems operate at several hundred volts. Since there are no batteries or charge controller to content with, the inverter and the PV array can get down to serious business.

We opted for a 24-volt system years ago, when we knew a lot less about this business than we do now. It was a choice that has worked well for us, though I must confess our array works more efficiently since we wired it for 48-volt operation using OutBack's MX60 charge controller (*see Chapter 8 for details*).

MICK'S MUSINGS

Pound for pound, dogs are twice as energy-efficient as cats. If this sounds high, calculate a feed bill for 50 pounds of cats.

How Much Generating Capacity?

Our home runs with power from the sun and wind (the Appendix tells you exactly what we have). Most of our neighbors are strictly solar. Maybe you'll be lucky enough to add micro-hydro power to your equation. After you read the next three chapters about solar/wind/hydro equipment, you'll have a better idea of what will work for you. You'll also want to study the detailed Appendix with worksheets, references, and resources.

How Many Batteries?

Note: *A Battery-Sizing Worksheet is provided on page 266. See also Chapter 9 - Batteries.*

While one of the most common problems with off-the-grid renewable energy systems is too few batteries, it is also possible to have too many. Why? Because it's not always enough just to keep the batteries in a so-so state of charge. The batteries really should be brought to a full charge

Our 6-volt batteries are enclosed in a box with a removeable lid for easy access and maintenance. The shunt (upper right) connects to the Trimetric meter.

regularly, especially if you are using a meter to monitor the batteries' state of charge, since most meters begin to lose their calibration after a few days unless the batteries are charged up. More than that, however, lead-acid batteries need to be equalized (over-charged, under controlled conditions) every so often to ensure that the plates remain free of sulfates. (Many people do this with a generator, but you shouldn't have to with a properly-sized system. Besides, it's cheating.)

The point is, your battery bank needs to be sized in accordance with your loads and your generating capabilities. With too few batteries you will be wasting sunlight and wind; with too many, the batteries themselves will suffer.

If you size your system perfectly, the first time out of the gate, then your success is attributable as much to luck, as to math. The key to sizing your system, then, is to leave yourself room to add on later. This means starting with the right inverter. It means running heavy enough wire to the array that you can later add more modules, without having to worry about line loss. And it means leaving yourself space in the electrical room—and the battery box—to add more batteries, if need be.

How Many Watt-Hours in a Battery?

To convert a battery's amp-hour capacity to watt-hours:

amp-hours x voltage = watt-hours

Example: 390 amp-hours (L-16 battery) x 6 volts = 2,340 watt-hours
Discharge to 50%: 2,340 x .50 = 1,170 watt-hours available

See Battery Sizing Worksheet on page 266.

WILLIE'S WARPED WITTICISMS

A Cat's Power Formula:
The power of a cat is directly proportional to the quality of its food, and inversely proportional to the quickness of the dogs it's forced to live with.

chapter 5

Solar Photovoltaic Modules

how to capture a ray of sunshine

The first time I ever saw a polycrystalline solar module from 2 feet away, I had the feeling I was looking at something magical. Deep inside, it looked like thin slices of the finest Persian lapis, overlaid with semi-transparent crystalline mirrors, tinted in a hundred hues between blue and purple. It might have been a master jeweler's creation, it was so awe-inspiring, except that closer to the surface lay a thin grid of aluminum channels; pathways for the energy with which this strange creation resonated. Were such a rectangular jewel of modern alchemy to travel back through time, I mused—to ancient Sumer or Babylon, perhaps—it might have changed the history of early civilization. Certainly it holds the power to change the history of *this* civilization, if only we, desensitized as we are to the finer facets of technology, could feel the pure wonder of what these sparkling blue panels are capable of.

Kyocera 187-watt solar module.
Photo by Kyocera Solar.

Until then, a minority of us will continue to eschew fossil fuels, in favor of sunshine—one bright ray at a time.

No one ever intended solar modules to be so beautiful, of course; they just turned out that way as a direct result of their design. Made from the basest of materials—nothing more special than thin layers of silicon (the stuff beaches are made of), coated (doped) on either side with atoms of dissimilar electrical properties—they take on their other-worldly appearance and their special properties through a laborious and painstakingly precise manufacturing process. The end result is a multi-layered substrate through which electrons are set in motion, after being jostled by sunlight.

The phenomenon of converting sunlight into electrical energy was first observed more than two decades before the outbreak of the Civil War, but it wasn't until the 1960s and the Cold War that the first efficient solar cells were produced for use in the space industry. Though exceedingly expensive, by 1970 more solar modules were produced for use on Terra Firma than in space. Since then, the use of solar modules has steadily increased, while the price has been driven down, from Buck Rogers' price of over $1000 per watt, to less than $5 per watt for contemporary earthlings. Still a little pricey, true, but no longer out of reach.

Types of Solar Modules

The primary energy producing unit of any photovoltaic system is the cell. Crystalline, or polycrystalline, silicon cells each produce about .50 volts (a little less for amorphous silicon, which will be described in more detail below), regardless of how large or small they may be. Bigger cells produce more amperage, but the voltage remains the same.

To increase the voltage to the point where it can be used to charge a battery, several cells are connected in series, and arranged together within a module. Also commonly known as a "solar panel," a module is one of the glass-covered, aluminum-framed units you buy from the dealer and install on or near your house.

Most residential modules on the market today range from 16.5

to 18 volts. This may seem high for a 12-volt system but, like water running downhill, it's necessary to keep the current flowing from the module to the battery, rather than the other way around. Solar cells experience a drop in voltage with increased temperature (during the summer months, for instance), and batteries in a 12-volt system, while equalizing, often reach 15.5 volts or more. If the module voltage were any less, there wouldn't be enough electrical "pressure" to bring the batteries to a full state of charge.

Modules are rated under optimal laboratory conditions—direct midday sunlight, at a cell temperature of 25 degrees Celsius (77 degrees Fahrenheit). Performance drops off with a change of sun angle, a rise in temperature (each 1 degree Celsius rise in temperature will cause a 0.5 percent drop in voltage) or, of course, the appearance of clouds. Common sizes for modules in residential arrays range from 40 to 150 watts, though larger modules are available. While it would seem that you would be able to get more watts per dollar by buying bigger modules, it's not always true. Prices fluctuate from dealer to dealer, week to week. If you're not too picky about size, you can usually find some good deals. (If you do buy smaller modules, however, you need to realize that they will entail more wiring per kilowatt, as well as more work and material for the mounting frame.)

Most solar modules in use today are made from crystalline, or polycrystalline silicon, protected by a layer of

---

Actual Module Output vs. Rated Output

The watt rating (Pmax) of a solar module is the product of its rated amperage (Ipmax) and its rated operating voltage (Vpmax). A 120-watt module might produce 7.10 amps of current at 16.9 volts (7.10 x 16.9 = 119.99). But when the batteries draw the voltage down to 12 or 13 volts there is not a corresponding rise in amperage, and a significant amount of power is lost (7.10 x 13 = a mere 92.3 watts). How do you get it back? Power point tracking can help. For a thorough discussion of this watt-saving technology, read the section on MPPT Technology, Chapter 8.

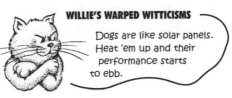

WILLIE'S WARPED WITTICISMS

Dogs are like solar panels. Heat 'em up and their performance starts to ebb.

tempered glass. They work well for most people in most applications, but they are not without certain shortcomings. As I mentioned above, the voltage begins to dwindle as the cells heat up. In addition, performance drops off considerably under conditions of partial shading, as often happens when snow slides off one half of a module, but not the other. In this case, the clear side tries to run current through the shaded side, to no avail. Since the module voltage is the sum of all the individual cell voltages, a partially shaded module may not have sufficient voltage to charge a battery. The same thing occurs when one of two modules wired in series is shaded. In systems over 24 volts, damage can occur to the shaded module when the unshaded module tries to drive it. For this reason, module manufacturers suggest the installation of bypass diodes to avoid damage when operating at higher voltages.

All of these problems have been overcome with the introduction of thin film, or amorphous, silicon modules. Utilizing the same technology as the solar cells on pocket calculators, these modules operate better than crystalline silicon in high heat, low light, and partial shading conditions. The only drawback—besides being less fascinating to the eye—is that they are less efficient, meaning that each thin film module needs about twice the surface area to produce the same wattage as an equally rated crystalline silicon module. Unless you are cramped for space, the extra size shouldn't be a problem. (You can figure, roughly, that a thin film module will produce 5.85 watts per square foot, compared to roughly 10.5 to 12 watts per square foot for crystalline silicon modules. A 1,200-watt array of thin film modules, then, would take up over 200 square feet of space, while a similar array with standard modules would require only a little more than 100 square feet. In terms of dollars per watt, the two types of panels are quite similar.)

UNI-SOLAR® products include flexible and portable modules, framed solar panels, as well as solar shingles that can replace traditional roofing materials. *Photo courtesy of Uni-Solar.*

Flexibility is an added advantage of some thin film silicon production methods. This makes thin-film silicon ideal for certain roofing applications. Uni-Solar®, in particular, has a developed a full line of thin film photovoltaic products that can be used either in conjunction with standard roofing, or in place of it.

Finding a Place for the Array

While it would seem logical that the roof would be the best place to locate your array, there are some (three, to be exact) drawbacks to rooftop mounts that you should consider. First, it's difficult to sweep off the snow. In the winter, when the mercury hovers around zero the day after a snowstorm, the snow may cling to the array all day, even if it's bright and sunny. You could be charging at full power, but you won't get any power at all if your modules are buried under 6 inches of snow. Even on days when it's foggy following a storm, there's lots of power to be had, as long as you can remove the snow.

Second, rooftop-mounted modules are hard to clean in the summer, after a little

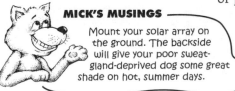

MICK'S MUSINGS

Mount your solar array on the ground. The backside will give your poor sweat-gland-deprived dog some great shade on hot, summer days.

Where is North?

Those of us deficient in surveying experience generally rely on a compass and a chart giving the magnetic declination for our particular area to find true north. If you're really lucky, you might be able to get within a few degrees by this method. If that is acceptable, great. After all, experience has shown that plus or minus 5 degrees in the alignment of the array makes virtually no difference in the output of power.

But why settle for an approximation with a fractious compass when you can get a nearly perfect alignment with a couple of T-posts and the ability to locate two of the most conspicuous constellations (the Little Dipper and the Big Dipper) in the night sky? Polaris, the bright, terminal jewel in the handle of the Little Dipper (officially known as Ursa Minor) is the North Star. It is the pivot point around which every other star in the night sky revolves. The two bright stars delineating the outer bowl of the Big Dipper (Ursa Major) point to it from 30 degrees away.

Once you know where Polaris is, all you need to do is drive a post into the ground near where the array will be situated. Then, as you sight directly over the top of the first post, have someone to the north of you move a second post east and west until a sight line along both posts points directly to Polaris. You now have a true north-south axis to work from for a perfect alignment of your solar array.

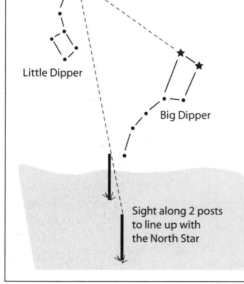

North Star

Little Dipper

Big Dipper

Sight along 2 posts
to line up with
the North Star

Due to a phenomenon known as the precession of the equinoxes, the North Star changes every few thousand years. In the unlikely event this book is still in print 12,000 years from now, some future editor will have to replace Polaris with the 0.1 magnitude star, Vega.

Cleaning snow off the ground-mounted solar array is easy.

sprinkle leaves a hundred dust particles, for every drop of rain. And third, it's a lot more work to adjust the array for seasonal variations in the sun's angle.

That being said, not everyone has a big, sunny yard or a south-facing deck, so for you the roof may be the only option. Fortunately, roof-mounted arrays no longer need to be the tedious and potentially leaky affairs they were in years past, since now commercially available standoffs can be painlessly incorporated into your home's roofing system. While the standoffs themselves do in fact penetrate the roof deck, they are made waterproof with standard plumbing vent jacks. Thus they will hold your array safely off the roof while virtually negating the possibility of leakage. For standing seam metal roofs, non-penetrating mounts can be used. Designed to work with stout, easily adjustable aluminum frames, standoffs take the guesswork out of roof-mounted arrays.

Whether on the roof or on the ground, the array needs to be as close to the house as possible, in a

Curt and Kelly's roof-mounted array.

Ed and Val's deck-mounted solar array.

Dave's custom-made rack holds six modules.

location where it can receive full, unimpeded sunlight for the 3 hours on either side of noon. Bushes or trees that cast a shadow against any part of the array can greatly reduce the energy output, even in winter when the branches are bare. You should either remove the foliage, or find a more suitable spot to place the array, even if that means the roof.

Arrays should always point due south, as long as there are no peripheral obstructions. Trees or hills to one side of the array may make it more advantageous to position the array at an angle that makes the most use of what sunlight there is.

If the best spot to locate the array is farther from the

John's pole-mounted array tilts for easy, seasonal adjustment.

house than 50 feet, you might consider designing your system to operate at 48 volts, since this will greatly decrease the diameter of the wire needed to carry current to the house with a minimum loss of current through the line.

Tilt Angle of the Array

The sun passes through 47 degrees of arc twice a year as it treks from the Tropic of Capricorn to the Tropic of Cancer, and back again. At 40 degrees north latitude (as we are here, in northern Colorado) the sun moves from 26.5 degrees above the horizon at noon on the winter solstice (December 21st), to 73.5 degrees at noon on the summer solstice (June 21st). Obviously, a solar array that is perfectly perpendicular to the sun will produce the most power. If you diligently adjusted your array every few days, you would want to set it at 90 degrees minus the sun's angle. The array on the first day of winter would be set at 63.5 degrees (90 minus 26.5 degrees), and gradually adjusted to the first day of summer at 16.5 degrees (90 minus 73.5 degrees).

But you won't adjust the array angle every few days, of course, unless it is tantalizingly easy to do so. (Keep reading...)

Shortly before we handed our electrician his walking papers, he brought a solar installer to our house in hopes of getting some advice on how to wire our system to code. The installer walked down to our cabin to check out my handiwork with that basic system. When he saw the array there set at 40 degrees, he told me the angle was way too low— we should set it at 55 degrees and leave it there. I thought the man was daft. Then I realized that his logic was shared by almost everyone around here.

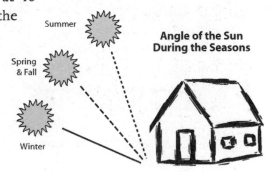

Summer

Spring & Fall

Winter

Angle of the Sun During the Seasons

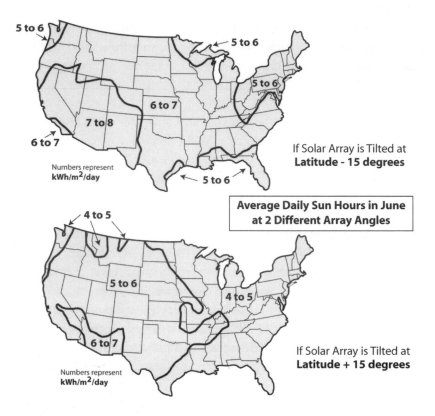

5 to 6
5 to 6
5 to 6
6 to 7
7 to 8
6 to 7

Numbers represent
kWh/m²/day

If Solar Array is Tilted at
Latitude - 15 degrees

5 to 6

| Average Daily Sun Hours in June |
| at 2 Different Array Angles |

4 to 5
5 to 6
4 to 5
6 to 7

Numbers represent
kWh/m²/day

If Solar Array is Tilted at
Latitude + 15 degrees

Should you tilt your array for the seasons?

These two maps illustrate the difference in sun hours if your array is tilted to the optimum angle (top map) versus if it is kept at the winter setting (bottom map). Data from the National Renewable Energy Laboratory Resource Assessment Program.

This is how it goes: if you set your array to capture the most light during the coldest, darkest time of the year, then whatever percentage of energy you lose in warmer months will be more than compensated for by the extra hours of sunlight.

This line of reasoning is based on the assumption that cloud patterns are essentially the same throughout the year, but they're not. In Colorado, for instance, the sky is clearer in winter. In summer, though it doesn't often rain, the clouds roll in over the western mountains practically every afternoon, greatly diminishing the amount of light that falls on the solar array. Couple this with the fact that sun rises and sets so far north in the summer that the modules need a low angle to capture any early and late day sun, and the argument for a steep, stationary angle falls apart.

That being said, I have to admit that most of the dozen or so people we know who live entirely off the grid have their arrays set at a steep angle, and never bother to adjust them for the seasons. It's probably the reason why LaVonne and I are often treated to a chorus of generators on warm summer nights, while our generator is resting comfortably in the garage. The bottom line? Make or buy an adjustable frame. Then you at least have the option, whether you use it or not.

■ Reckoning Where The Sun's Going To Be

If you choose to adjust the angle of your array (and you should, considering how much money you're paying for it), it's very simple to measure the sun's angle. All you need is something that casts a shadow (like a deck railing, a fence post, or even the backside of your solar array), a long, straight board, and a floating-pointer angle finder that you can buy at any lumber yard or hardware store. At midday, when the sun casts a shadow directly north, place one end of the board on top of the post (or railing), with the other end resting at the end of the post's shadow (on flat ground, of course). Place the angle finder flat on the top edge of the board, and it will show you the sun's angle; turn it 90 degrees on its side, and it will show you the optimum angle for your array.

Place the angle finder along a board at high noon to find the sun's angle.

Interestingly, if you take a measurement at midday on the spring or fall equinox, you will find the sun's angle to be 90 degrees, minus latitude. At 40 degrees north latitude, for instance, the sun will be at 50 degrees on the equinoxes, and will change 23.5 degrees up or down as the seasons change.

Using this information, most people who do adjust their arrays do it four times a year, going from latitude plus 15 degrees in winter;

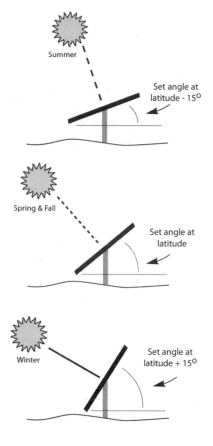

Seasonal Tilt of the Solar Array

to latitude in spring; to latitude minus 15 degrees in summer; and back to latitude in the fall. These seasonal adjustments give you the average optimal amount of sunlight for each period.

If you design your array with these three adjustment angles, you need to realize that the summer and winter angles will stay in place longer than the spring and fall angles. In other words, your summer and winter settings will last for about four months each; your spring and fall settings for about two months. If the sun angle is between two adjustments, opt for the lower setting; you'll capture more sunlight. (If you want to improve on the 4-times a year adjustment strategy, see the next section for an easy and incremental method of adjustment.)

Mounting the Modules

We began our solar venture with six 110-watt Solarex modules. Lacking the keen edge of experience, I built a massive frame to hold the 72 square feet of modules. The exterior frame, as well as all the inside rails, were made with 2 x 3/16-inch angle iron. It weighed about as much as a big full-grown man, and it took four men to carry it down the steep, rocky slope in front of the cabin to mount it on the concrete piers we had poured a week before. (As unpleasant as that may have been, it was a picnic compared to the chore of carrying the frame back up the hill when we moved it to the log house.)

Being several degrees wiser when it came time to mount the four 120-watt Kyocera modules we later added to the array (and another four, even later), I used 1½ x ⅛-inch angle iron for the outside frame, and 1½ x ⅛-inch strap

The back side of a pole-mounted PV array shows UniRac's mounting system, and the black junction boxes of the solar modules. *Photo courtesy of UniRac.*

iron for the inside rails. The finished frame weighed—and cost—about a third what the first frame did. It was easy to mount, and affords all the support the modules will ever need.

While commercially manufactured, pole-mounted aluminum frames are available and widely used, you may be better off building, or having them built, yourself. That way, you can engineer them to fit the terrain, and to adjust in the way you would like them to.

The bottom sides of our frames pivot on heavy iron supports set in concrete. For adjustment, we originally used three sets of legs for the top sides (one set for winter, one for summer, and another for spring and fall). The only problem with this arrangement was that the panels and their frames are quite heavy, making seasonal adjustments a

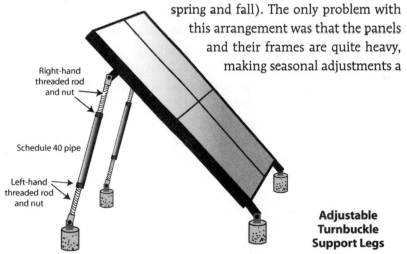

Right-hand threaded rod and nut

Schedule 40 pipe

Left-hand threaded rod and nut

Adjustable Turnbuckle Support Legs

dicey two- or three-person operation. Since then, we have replaced the old rigid legs with two sets of adjustable turnbuckles I made from right- and left-handed 1-inch threaded rod, with right- and left-handed nuts welded to either end of a length of sturdy pipe (the buckle). The first set of legs adjusts from 62 to 40 degrees, the second set from 40 to 20 degrees. The legs are easy to change out because the array does not have to be raised or lowered during the process. Best of all, I only have to give each buckle a few turns every so often to keep the array at the optimum angle, year-round.

Another way to minimize heavy lifting when changing the array angle is to buy (or design) a mount where the frames pivot in the middle. Centering the frames on one or two heavy steel pipe posts is an ideal solution, providing you can set the posts deep enough in the ground—and with a broad enough concrete base—to prevent the whole array from toppling over in high winds. (Since bedrock is less than a foot down where we set our array, we'd have been tempting fate by mounting the frames on a pair of central posts.)

■ Trackers

Commercial mounts called trackers are available for anyone who wants to maximize the potential of their solar array. Trackers move the array in step with the sun as it travels from east to west across the horizon. Conventional wisdom holds that trackers are a good investment in the sun belt, but not in higher latitudes where the sun rises and sets so far south during the winter months that the small amount of extra sunlight wouldn't be worth the extra expense, or the hassle of having an added component that might occasionally require service or repair. For most systems, it makes more sense to just buy extra modules since things that don't move are generally less headache than things that do.

Wiring the Array

While most common modules are wired for 12-volt operation, many larger modules will be wired for 24 volts from the factory, or can be changed from 12 to 24 volts. All the modules you buy will come with

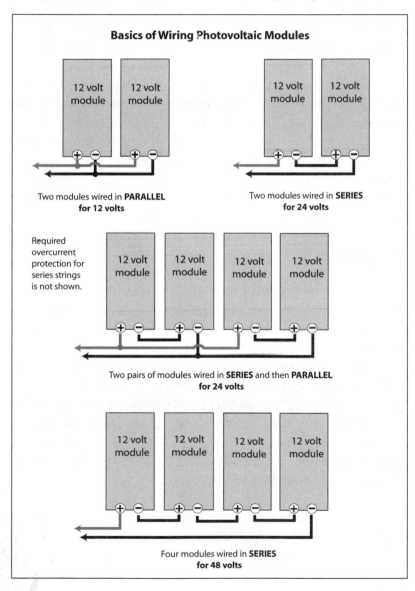

Basics of Wiring Photovoltaic Modules

Two modules wired in **PARALLEL**
for 12 volts

Two modules wired in **SERIES**
for 24 volts

Required overcurrent protection for series strings is not shown.

Two pairs of modules wired in **SERIES** and then **PARALLEL**
for 24 volts

Four modules wired in **SERIES**
for 48 volts

diagrams detailing how to wire them in parallel and in series. Since every brand of module is set up a little differently, there's no point diving into specifics here. A run through the basics, however, would be helpful before you order the modules.

In a 12-volt system, all the modules are wired in parallel—positive to positive, negative to negative. Thus, the amperage of each individual module adds to the total amperage of the array, without increasing the voltage.

In a 24-volt system, each pair of modules is wired in series—positive to negative, and vice versa—doubling the voltage (from 12 to 24). Then each series string is wired in parallel to increase the amperage. (Likewise, in 48-volt systems, sets of four modules are wired in series.)

The important point to remember is this: only modules of identical wattage should be wired in a series string, since the amperage of the string will be equal to the amperage of the weakest module.

■ Fuses and Breakers

Beginning with the fragile aluminum channels on the surface of the individual solar cells, and ending with the heavy copper leads going to the charge controller, the wiring of a solar array is like a complex watershed. Hundreds of tiny tributaries flow into dozens of larger ones, which in turn dump their current into a few even bigger channels, before emptying into the great river that ends at the load. (It doesn't exactly end at the load, since there is a pathway back to the source, but that's another story.)

As long as the integrity of the system is intact, the flow of current is orderly. But when a short circuit occurs somewhere along the line, large amounts of current can be sent back in a reverse flow, with potentially disastrous results. It's like what would happen if a cataclysmic seismic event sent the waters of the Gulf of Mexico raging back up the Mississippi River into each of its progressively narrower tributaries: the small streams would hardly be able to contain the additional volume of water.

To keep a similar occurrence from taking place within the confines of your array wiring, it's important to install fuses and/or

OutBack combiner boxes.
Photo courtesy of OutBack Power Systems

circuit breakers wherever amperage is combined. Usually, the leads from each series of modules terminate into a combiner box, where the individual positive leads are directed through fuses or breakers of appropriate size. To reduce the risk of a costly mishap while working on the array wiring (and to satisfy the electrical inspector), the main lead coming from the combiner box is then run through a circuit breaker before going to the charge controller.

■ **Grounding Your Solar Array**

You should pound a ground rod into the earth next to the array, and then run a heavy copper wire from it to the common house ground. Also, you will have to be sure that each frame, and each module within the frame, has a path to ground. To be certain, you can use a multimeter to run a continuity test from the ground wire to the frame of every module.

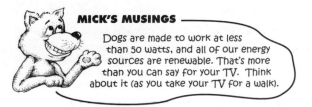

MICK'S MUSINGS

Dogs are made to work at less than 50 watts, and all of our energy sources are renewable. That's more than you can say for your TV. Think about it (as you take your TV for a walk).

Leaving Room to Grow

Even with the most assiduous planning, there is always a certain degree of guesswork involved in determining how many watts of solar power you will need. You don't want to be running the generator every time you wash a load of clothes, but you also don't want to ravage your kids' college fund to buy modules you don't need.

The best answer is to start with what you *know* you'll need, and then add more modules after you've had a taste of what the system is capable of. This is especially true if you are including a wind generator in the equation. Aside from ensuring that you have adequate space (on the ground, on the roof, or in front of the deck) all you'll really need to do when you set up the initial system is to run heavy enough wire to handle the amperage of any system you might conceivably build, or run an extra conduit in the trench. It will save a lot of digging later. Then, when you look at your array, rather than thinking of work, you can admire its inherent beauty.

John's second pole-mounted solar array with four modules, and room for four more.

How Do Solar Cells Really Work?

Largely because of the ubiquity of computers, anyone who has not been living in a cave for the past two decades knows that silicon is a semiconductor of electricity—it sort of allows an electrical current to pass through it, but hardly with the facility of copper or aluminum. Oddly enough, it's this quasi-standoffish attitude of silicon that makes it so useful in the manufacture of solar cells.

Crystalline silicon solar cell.

   Chemically, silicon has 14 positively-charged protons, and 14 negatively-charged electrons. This would seem to be a happy arrangement, if not for the fact that it has room for four more electrons in its outer energy level. How does it get them? It could snatch four passing electrons from somewhere, but there would be no protons to hold them in place, so the kidnapped electrons would soon escape. So instead, it borrows them from other silicon atoms, forming a crystal lattice in the process (except in the case of amorphous silicon). In this crystal, every atom of silicon is attached to four other atoms of silicon and they all share electrons. In other words, every silicon atom has the four extra electrons it wants, with no net charge, since the protons in the crystal exactly balance out the electrons. It's a really cushy setup.

Silicon Doping: Homogeneity's Undoing

In fact, it's far too cushy for our purposes. Happy silicon with happy electrons is pretty useless if we want it to do any work. We need to stir things up a bit. How? By adding impurities to it. Say, a few atoms of phosphorus. Phosphorus has five electrons in its outer energy level, so if it is introduced into the silicon crystal lattice (in a process called doping), that fifth electron will be frantically looking for a place to fit in. Now we have an unhappy electron, and that makes us happy.

   But we're just getting started. A melancholy electron wandering aimlessly in search of a home doesn't do us much good. We need to give this electron a purpose—something it can aspire to. We do this by doping the other side of the silicon crystal, this time with a different impurity, say boron. Having only three electrons in its outer energy level, a boron-doped silicon crystal will have empty spaces where electrons could be, but aren't. These empty spaces are called holes, and each of these holes would like to have an electron to call its own.

Are you beginning to see where this is leading?

Our phosphorus-doped silicon is called n-type, in honor of the extra negative electrons, and the boron-doped silicon is called p-type for the extra positive holes. And we're getting very close to having a useful electronic device.

Life at the P-N Junction

At first blush you might think that all the extra electrons in the n-type silicon would zoom across the p-n junction (the place where the two opposite types of silicon meet) and fill in all the holes in the p-type silicon, but it just doesn't happen that way. Oh, a lot of them start out fast enough, but quickly begin to have second thoughts. Sure, an electron soon realizes, there may be a nice cozy place for me on the p-side, but my faithful proton is still on the n-side. I'm so confused. It's a bit like young love.

The important thing to remember is that, even though n-type and p-type silicon have extra electrons and holes, neither type, alone, has any net electric charge. In both cases there are just enough electrons to balance out the protons. But once the rush across the border occurs, that quickly changes. Every time an n-type electron jumps through the p-n junction and fills in a p-type hole, a negative charge is created on the p-side, while a positive charge springs up on the n-side in the place where the electron was, but no longer is. Once everything settles down, we find that there is a great gathering at the p-n junction, with negative electrons lining up along the p-side, and positive holes lining up along the n-side. This creates an electrical equilibrium, and if we left things like that the p-n junction would be a really boring place.

But we're not through yet, for now it's time to finish building our solar cell. To do this, we need to crisscross the surfaces of

Closeup of a polycrystalline solar module.

our silicon wafer with electrically-conducting channels. This will provide an easy path for the electrons to travel along, once we add the magical ingredient, sunlight.

When a photon of light of the right energy and wavelength strikes an electron hanging out with all of his buddies on the p-side of the p-n junction, the electron is instantly imbued with a jolt of energy and is suddenly free to move around. Where does it go? It can't go any farther into the p-side; there's quite a crowd there already. So instead it uses all this free energy to make a beeline back to the n-side. And, with a little luck, it will be picked up by one of the conductors on the surface of the n-layer and sent through an electrical circuit.

A Loopy Idea

Once the process begins, the electrical equilibrium at the p-n junction is hopelessly undone and the proverbial floodgates are opened. In an instant, multitudes of electrons that were just moments before hanging out at the p-n junction, are whisked out of their silicon Shangri-la, drawn through the windings of a washing machine motor or the filament of a light bulb, and unceremoniously dumped back on the p-side of the solar cell, totally exhausted. But, like battered and beaten heroes in a video game, all they really need is a little nourishment—a single photon—to be right back in the thick of things.

The completed circuit is the key to making the whole thing work. Since a solar cell acts as a diode—only letting the current flow from the p-side to the n-side—it wouldn't produce electricity for very long without a fresh supply of electrons continually re-entering the solar cell from the p-side. That's why all solar panels have positive and negative terminals. The electrons flow out of the negative terminal which conducts electrons from the n-type silicon, through the load (the above-mentioned washing machine or light bulb) and back into the p-type silicon via the positive terminal.

Personal Power Companies

Homeowners	Dan & Kim
# of Occupants	2 adults
Location	Vinton, Iowa
Home Size	1,600 square feet on 2 levels
Home Heating	Jøtul wood stove; baseboard electric heat disconnected
Water Heating	Propane on-demand; solar collectors
Grid-Tied?	Yes
PV Array	1,200 watts ground-mounted on a Zomeworks passive solar tracker, south of house
Charge Controller(s)	Solar Boost 50
Inverter(s)	Trace (Xantrex) SW4024 sine wave
Batteries	12 Absolyte IIP 2-volt maintenance-free; 610 amp hours @ 24-volts
Wind Turbine /Tower	Bergey 1,500 watt *(discontinued model)* on a free-standing, 87-foot lattice tower
Solar Hot Water	2 Solahart 3 x 5 collectors with 80-gallon tank
Backup Generator	none
Comments	Dan and Kim have lived comfortably with their RE systems since 1999, when they decided to make the switchover to renewable energy prior to Y2K. Since 2004 they have been net metering with their local Rural Electric Cooperative. Their PV/wind system generates, on average, 5 kWh per day (75% from the sun and 25% from the wind) and they use approximately 4 kWh to power their home. The remaining electricity is sold back to the local utility.

chapter 6

Wind Turbines

making good use of the stuff of clouds

It's early April as I write these words. The mercury hangs just above freezing and the skies are thickly overcast. Beyond my office door, a wind chime rings melodically. If I walk outside, I can see our 1,000-watt wind turbine working, just beyond a copse of juniper trees. Though I can only feel a light breeze from where I stand, 50 feet above me the propeller blades are spinning furiously. I judge it to be about a 6- or

7-amp wind. The gentle whirring of the blades is a comforting sound. It's the sound of nature's energy being refined and transformed.

To me, it's music.

For the two years we were building our house, the wind sorely tested our sanity. Every time I tried to drop a plumb bob, the wind came out of nowhere and deflected it. I'd throw a sheet of plywood over my shoulder on a perfectly calm day, and instantly I'd feel a playful breeze, trying to wrest the sheet from my grasp. I'd walk up a ladder with it, and the breeze would grow stiffer with every rung. If I moved the ladder to the other side of the house, the wind would change direction and follow .

After awhile I became convinced that I could control the speed and direction of the wind, just by how I selected my activities. And who's to say I'm wrong? Not me; I'm a believer.

But, even if a person can—by using plumb bobs and plywood—make the wind do a few parlor tricks, it's a fact of nature that you will never be able to outsmart it. Yelling at the wind, or trying to ignore it, just encourages it. Pleading with it will elicit no sympathy, whatsoever. So you might as well face the facts: you'll never get along with the wind until you change the nature of your relationship.

Bergey's 7.5 to 10 kW Excel turbine. *Photo courtesy of Bergey Windpower.*

Having put up with the wind through the entire construction of our house, I couldn't wait to hook up the wind turbine and watch the wind do some good for a change. After we laboriously erected our tower and installed and wired the turbine, we had the pleasure of watching the propeller spin in the light breeze for about five minutes—before stopping altogether.

For two full days the blades didn't make a single revolution. By the end of the second day I began thinking that buying a wind turbine was the stupidest thing I'd ever done. What was I thinking? I could've bought 500 watts of easy-to-install solar modules for what I paid for one obstinate machine, sneering at me from 50 feet in the sky.

It was almost as if the wind wanted to know it would be appreciated, before it was willing to do any work. And appreciate it I did, when it finally began to blow again after two days' absence. I believe it was the first time in my life I was actually happy to feel the wind. Obligingly, it blew more or less steady for the next several days, before settling back into its old, unfathomable non-patterns.

Since that rocky start two years ago, when I wasn't at all certain we'd made a good investment, I've begun to learn just how important the wind factor is in our energy equation. Though predicting the strength and duration of the wind at any given moment will never be more than a guess, over time the wind always comes through for us.

Would I now trade our 1,000-watt wind turbine for 500 watts of solar modules? Not a chance. The wind provides us with power when the sun can't, which means that we can get by with fewer batteries and less worry. Our wind turbine usually doesn't supply all the power we use during a stretch of cloudy weather, but it often provides enough to get us by. Then, when the sun finally does show itself—escorted by a breezy change in atmospheric pressure—the batteries recharge all that much faster.

Wind power is not for everyone, of course. In many areas of the U.S. the force of the wind near ground level is too slight to be of practical value for a home-based wind system. And even if you do have usable wind power where you live, other factors need to be considered before you rush out and buy a wind system.

A Matter of Geography

In addition to the wind itself, you will need to have enough land to safely locate the tower on which the turbine will be placed. Usually, this means an acre or more. Some folks are able to avoid installing a tower by mounting one or more small wind turbines (500 watts or less) on a barn or other non-living outbuilding, where the vibration caused by the spinning blades will not shake the plaster off the walls and induce periodic fits of insanity. However, for larger machines (600 watts and up), a tower is a must.

There are two things to keep in mind when searching out a spot to place a tower. First, it should be far enough away from living areas and property lines that no one would be injured (or worse) if the tower fell over, or if all or part of the wind turbine came flying off in a killer wind or microburst—worst-case scenario stuff, in other words. How far is far enough? A good rule of thumb is 15 rotor diameters away from the house. So if your turbine has, say, an 11-foot propeller, you'll want to place your tower at least 165 feet away.

Secondly, the tower should be high enough to clear any obstacles that might be in the path of the wind. Ideally, the turbine should be mounted at least 30 feet above the tallest object—tree, building, hill, etc.—within 300 feet. This requirement is to minimize the effects of air turbulence, which is to a wind turbine what a washboard road is to a car. Besides causing extra wear and tear on the turbine, turbulence greatly diminishes the force of the wind.

How Much Wind is Enough?

If you have the land with an ideal location for a wind turbine and you've determined that neither your neighbors nor the local bureaucrats have any objections to a tower, you'll need to decide if you have enough wind at your site to justify the time and expense of installing a wind system. As you set out on this quest, the first thing you will discover is that exact wind data for your particular location is probably non-existent, unless your home is next to an airport or a military base. But you can still get a pretty good idea what the force of the wind is in your area by referring to the wind maps for your state. The Department of Energy (DOE) maps at the Bergey Windpower website (*www.bergey.com*) assign a wind class number to every square inch of every state. Though these maps are painted with a rather broad brush, they still offer a lot of insight for the wind resources that are available in your area. For further clarification, you should also read the National Renewable Energy Laboratory's Wind Energy Resource Atlas of the United States at *http://rredc.nrel.gov/wind/pubs/atlas/chp1.html*. This is a thorough document that discusses national and regional wind patterns, seasonal variations, and the painstaking methods used to compile the data.

A quick glance at the national map *(see page 262)* will show you that the most paltry wind resources are in the Southeast, while large areas of excellent wind are in the upper Midwest, particularly the

Southwest Windpower's Whisper 100 (the re-engineering H-40). *Photo courtesy of Southwest Windpower.*

Dakotas and the western edge of Minnesota. Good winds can also be found in the higher terrain of both the Northeast and Northwest, and all along the Rocky Mountains.

You might also want to view the maps at the Southwest Windpower site (*www.windenergy.com*). Southwest has collected links to the best maps available for each state. Some were compiled by NREL, others by TrueWind Solutions. In some states you will be able to input geographical coordinates and print out an extensive data sheet showing, among other things, the average strength of the wind from sixteen different directions, as well as the average wind speed and power density during different seasons and from varying heights above the earth's surface. Bear in mind, however, that this information results from extrapolation of existing data from the nearest sites where measurements actually were taken. No one really knows for sure what the wind characteristics are at the top of the tower you haven't erected yet.

To be absolutely sure there's enough wind at your site, you could buy an anemometer and monitor its readings for a few months. If you get a fancy recording anemometer, or one with a computer interface, you will actually be able to plot wind patterns over time to determine what the average wind speed is at your location for different months of the year. Should you go this route, however, please keep in mind that your anemometer readings will not be entirely accurate unless you are able to mount the instrument at the same height as your proposed wind turbine. The lower you place it, the less encouraging the results.

To make things easier, the folks at Iowa State University have compiled a Wind Energy Manual that will tell you, among other things, where best to locate your anemometer and how to extrapolate the data you collect to calculate probable wind speed at different heights above different types of terrain. You can download this handy document at: *www.energy.iastate.edu/renewable/wind/wem/wem 08_power.html.*

But you really shouldn't have to sort through mountains of data, or spend a lot of money on wind monitoring equipment, for the simple fact is, if you think you have enough wind at your site, in all likelihood you do. As I mentioned in Chapter 3, my own personal rule of thumb goes as follows:

If the wind blows often enough and hard enough to annoy you, you can probably make good use of a wind turbine.

However you go about it, there are some surprising facts about wind speed and the amount of power you can hope to harvest from the wind. For starters, the relationship between the speed of the wind and the power it generates is not a simple linear correlation. What am I talking about? Just this: a 30 mph wind is not, as you might imagine, half-again as powerful as a 20 mph wind—it's nearly 3.4 times stronger! How can this be? It's because the force of the wind increases as the cube of the wind speed. So, 20 x 20 x 20 = 8,000, while 30 x 30 x 30 = 27,000. If you then multiply either of these products by 0.05472, you will discover the force of the wind in watts per square

---

Wind Speed Conversion
meters per second (M/S) x 2.23 = miles per hour (mph)

---

meter (W/m^2) at sea level for that particular wind speed. This is a tidy arrangement, because it turns out that solar radiation is also measured in W/m^2, so it's a simple matter to compare the speed of the wind hitting the blades of a turbine, with the sunlight that falls on an array.

And how do they compare with one another? Generally, the power of the sunlight hitting the earth (or your solar array) in the middle part of a summer's day at mid-latitudes is equal to a steady wind speed of 22 to 23 mph—about 600 W/m^2.

This isn't the amount of power you'll be sending to your house, however. Your solar array will only be able to reap around 12 to 15 percent of this energy, and these figures hold fairly well for wind turbines, too, though efficiency percentage is not a commonly-used term with home-based wind turbines, owing to the fact that similarly-rated machines may have vastly different sweep areas.

However you measure it, it takes a pretty stiff breeze—square meter for square meter—to rival the power of the sun; far more wind than is blowing around in most locations. Considering the psychological effect wind has on a lot of people, this is probably a good thing. But it also makes your decision to install a wind system more diffi-

Our tower rises at least 25 feet above the trees.

cult since you might live in a fringe area, where there may or may not be enough wind to make the installation of a tower and turbine a successful venture.

If the average annual wind speed where you live is 10 mph or more, you can almost be assured of having enough wind to reap a useful bounty of power from the unsettled atmosphere. This is because, unlike a solar array, a wind turbine's capacity to produce power is not limited to the hours between sunrise and sunset—it can produce power day or night, rain or shine.

The DOE maps list wind speed in power classes from 1 to 7. The upper limit of Class 1 winds approach 10 mph, provided the turbine is mounted high enough above ground. If you live in a Class 1 area you should probably do some homework before opting for a wind system. Living in a Class 2 area, though more promising than Class 1, does not assure you of enough wind, especially if your site is in a valley, near the lee side of a hill, or surrounded by towering trees (unless you are able to raise your tower at least 30 feet above the tallest nearby trees). By contrast, hilltops, coast lines and high plains make excellent sites for gathering wind. In particular, mountainous regions with large fluctuations in altitude are cauldrons of atmospheric change. Even without significant variations in atmospheric pressure (fronts), warm air will rise from the valleys during the day, while cool air will flow back into them at night, providing usable power while the sun is absent. If you are fortunate enough to have a home above a valley, but below a distant ridge, you should have plenty of wind.

I cannot over-emphasize the importance of tower height in designing a wind system. There really is a lot more wind up there, and it will invariably be steadier and less turbulent than the gusty, chaotic breezes that occur closer to terra firma. In fact, the DOE generally considers the wind power density at 50 meters (164 feet) to be double what it is at 10 meters (33 feet). The actual figures will vary over different types of terrain, of course, but it's still an eye-opening exercise in mathematics. You probably won't erect a 164-foot tower, but the DOE figures do make a point: height is good (just like voltage).

Although most modern wind turbines begin to spin—and thereby produce some amount of power—at 6 to 7 mph (the "cut-in" wind speed), they will not really begin to produce much in the way of usable power below 8 or 9 mph. For battery-backup systems, this is enough wind to keep the batteries charged and help reduce your electric bill. If you instead opt for a direct grid-tied system, you'll want an annual average wind speed of at least 10 mph.

LaVonne and I never bothered to look at the DOE tables before buying

MICK'S MUSINGS

A good wind is one that makes Newt's ears stand up straight. A great wind sends a cat or two spinning off the side of the mountain.

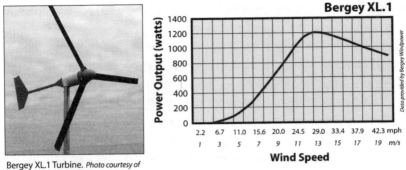

Bergey XL.1

Power Output (watts)

1400
1200
1000
800
600
400
200
0

2.2　6.7　11.0　15.6　20.0　24.5　29.0　33.4　37.9　42.3 mph
1　　3　　5　　7　　9　　11　　13　　15　　17　　19 m/s

Wind Speed

Data provided by Bergey Windpower

Bergey XL.1 Turbine. *Photo courtesy of Bergey WindPower.*

a wind turbine because at the time we didn't even know such data existed. Instead, we made a judgment based on the annoyance factor and it paid off. You'll have to decide which criteria will work best for you, but the above mentioned resources should go a long way toward guiding you to an informed decision.

Wind Turbine Basics

Take a lengthwise coil of wire and rotate it between two oppositely-charged magnetic poles, and you will produce alternating current. Spin it at the right speed (frequency), and you will produce a usable current for running AC motors and appliances. But try to charge a battery with this crude little generator, and it won't work. Somewhere along the line, the AC has to be converted into DC.

Some wind turbines—mostly (but not exclusively) those that are rated at 500 watts or more—send 3-phase AC through the lines and into a charge controller, where a series of rectifiers convert it into DC for storage in the batteries. Once it senses that the batteries are charged, the charge controller shunts excess power to a heat sink.

Other wind turbines do the AC to DC conversion within the turbine, itself, and send DC through the lines and into the batteries. Certain advanced models go a step further, incorporating electronics that sense the batteries' state of charge and adjust the rotor speed accordingly. On these models the propeller stops completely once the batteries become fully-charged, eliminating the need for both a separate charge controller and a heat sink.

Is AC better than DC? Not really. The system voltage is more important than the type of current the turbine produces. As we've already discussed, a wire that is rated to carry 40 amps of 12-volt current (480 watts) a distance of 50 feet, with 2 percent line loss, will carry 40 amps of 24-volt current (940 watts) 100 feet. In other words, by doubling the system voltage you can move twice the wattage twice the distance through the same wire, giving you a four-fold return on your investment. Bigger turbines, then, that need to be placed farther from the house, should operate at higher voltages. This means, of course, that your entire wind/PV system will have to operate at that same voltage. So if you plan to do wind, plan early.

While machines made for battery-based systems will generally be configured at the factory for 12-, 24- or 48-volt operation, turbines designed for direct grid-tied operation churn out high-voltage DC from the turbine (150 – 350 volts for the Windy Boy 1800U inverter, 250 – 550 volts for the 2500U) that is converted to grid-compatible AC within the inverter. For this reason, direct grid-tied wind systems are more efficient than battery-based systems—up to 94 percent efficient—since there is no energetically-expensive transformation from DC electrical energy, to chemical energy with-in a battery, and back again into electrical energy that must be further converted from DC to AC by the inverter.

Which system is right for you? What's true for solar is true for wind. If you live in the backwoods of Minnesota where blizzards and ice storms can tear down power lines in the blink of an eye and outages last for days on end, then you'd best be looking for a warm, cozy place to keep your batteries. On the other hand, if you live on the outskirts of a town with reliable grid power, then it might make more sense to opt for the more efficient direct grid-tied system.

Component-wise, there is very little difference between solar and wind systems. Wind/battery systems are set up just like solar/battery systems. In fact, many wind charge controllers have additional inputs for a solar array. And for direct grid-tied wind systems, the Windy Boy inverters from SMA America have the same features as the Sunny Boy and other comparable grid-tied solar inverters.

The main difference between solar and wind systems—beyond the obvious, of course—is that wind systems require more thought and more homework, owing to the plethora of different turbines and towers to choose from, as well as the widely-varying wind conditions from site to site. Talk directly with the manufacturers, not just the guy who wants to install the type of wind turbine he just happens to sell.

Will Wind Power Work For You?

This is a tough question that requires several regrettably slippery variables to be considered before answering. Where we live in Colorado, the sun shines over 300 days per year. I estimate the average wind speed at our house to be between 10 and 11 mph. Our current PV system is rated at 1,620 watts, our Whisper wind turbine at 1,000. I have not metered the wind output (yet), but I'd estimate that it is between 10 and 15 percent of our total energy production. Were we to purchase a different 1,000-watt machine designed to run optimally at lower wind speeds—a heavier turbine with a bigger sweep area—we could reap a lot more wind than we currently do. But any way we slice it or dice it, we're in a good wind area.

If you live in the Southwest where the sun shines high and often, wind will not be as accessible a resource as sunlight, and you'll need to do a bit of research before deciding if a wind system is worth the investment. But if you live in the Northern Great Plains, where the wind is fairly constant and the sun fairly isn't, then you will certainly want wind to figure prominently into your energy scheme.

Does anyone near you have a reasonably new wind turbine? Knock on their door and ask them about it. No amount of hair-pulling, divination, blind-guessing or rough-calculating can outdo solid experience.

Cost is always a consideration. Bigger turbines are not that much more expensive than smaller ones, but there are greater peripheral costs involved. Bigger means farther (from the house), higher (off the ground), and sturdier. While some small turbines can be mounted on the roof (provided the roof is built with standard trusses for mounting the mast), large turbines need to be mounted atop towers.

As I mentioned earlier, you can sidestep the tower issue completely by mounting two or more smaller turbines along the peak of the roof, or next to the wall, of a barn or workshop. You'll pay more per watt for the these smaller turbines, but make up the difference with what you'll save on the tower. The problem with this solution is that the turbines will be too low to the ground to take advantage of the really good winds, higher up.

As with any major purchase, you'll want to buy your turbine from a reputable manufacturer that stands by its products. Shop around and ask lots of pointed questions. Any salesperson worth his or her salt should be able to answer all of your questions about the various models and help you make an informed decision. Besides the pros and cons of different turbines, you will also want to know all the warranty details and how difficult it is to obtain warranty service. Who do you call with questions about installation and wiring? And who do you call once the unit is installed? Can you talk to a human without running through a computerized maze? It's good to know these things before you buy. After deciding that you can afford both the turbine and the tower to mount it on—and the extra wire—you'll need to consider how much work will be involved in getting the tower from the ground to the sky. Is it something you can do yourself, with careful planning, or will there be extra costs involved in erecting the tower?

If all this is beginning to sound like a lot of work, that's probably because it is. But just think of all the work you'll be getting out of your turbine, once it's up and running. In the long run it will all be worth it.

Turbines: A Quick Look at the Windy Beasts

When home-based wind turbines are discussed, the image conjured up in your mind is probably that of a horizontal-axis machine. Consisting of a propeller, a rotor, a generator, and usually a tail, these turbines resemble wingless aircraft with oversized propellers. Though they may all look somewhat similar from a distance, there's a lot of difference between turbines, and your success or failure as a wind farmer will largely depend on which one you choose. Different machines are designed for different types of wind. Generally, machines with large sweep areas, such as African Wind Power's AWP 3.6 and Southwest Windpower's Whisper 200 (the re-engineered H80), are engineered to operate optimally in lighter winds. Other machines, including the Whisper 100 (the re-engineered H40), have shorter propeller blades and are designed to take the punishment meted out at hilltop locations and during severe storms. Still other tur-bines, such as the machines produced by Bergey Windpower and Proven

An Air 403 turbine is mounted next to an exterior wall, and guyed to the roof. *Photo courtesy of Southwest Windpower.*

Energy, can endure some really nasty weather and still perform well in light winds.

Comparing wind turbines apples-for-apples will take a little research. Different machines have different cut-in speeds and different rated wind speeds, which is the speed at which optimal performance is achieved—usually in the 22- to 29-mph range. The table below lists the rated wind speeds, along with other pertinent data, for a few popular machines. It should be noted that there are many well-built turbines out there that aren't listed in the table. I'm not playing favorites; I just wanted to give you a good cross-section for the purpose of comparison.

Practically all turbines on the market today are 3-blade machines. The 3-blade design runs smoother than a 2-blade unit, and will be a little more efficient at converting wind into watts. As a general rule, the blades on smaller or lighter-duty machines are made from poly-propylene, while those on heavier machines are epoxy-coated wood or fiberglass. If damaging winds sweep across your site from time to time, you should avoid plastic blades on turbines of 1,000 watts or more. Trust me; this is the voice of experience talking.

A braking mechanism is also a handy feature, especially if you live in an area with ferocious gusts that could possibly damage your blades, or where ice storms might cover them with a layer of hoarfrost that can throw the system out of balance. With a wind brake—either mechanical or dynamic (electrical)—you can stop the turbine from spinning and wait for the sun to melt the frost or ice.

While the initial shock to your pocketbook will obviously be greater for larger turbines, the ratio of wattage gained for money spent will also be greater. A larger machine will also outlast one or more smaller ones, and a heavy, slow wind turbine will have a longer lifespan than a light, fast one. So, if you're going the way of the wind, buy as much as you can afford.

WILLIE'S WARPED WITTICISMS

A smart cat will always stay downwind of the dog, no matter how unpleasant the smell.

Comparison of Popular Wind Turbines

	Proven Engineering WT600	Bergey Windpower XL.1	SouthWest Windpower Whisper 200	African Wind Power 3.6
Rated Power	600 watts	1.0 kW	1.0 kW	1.0 kW (DC) 1.6 kW (AC)
Cut-in wind speed	6.0 mph	5.6 mph	7.0 mph	6.0 mph
Rated wind speed	26 mph	24.6 mph	26 mph	25 mph
RPM @ rated output	500 rpm	490 rpm	900 rpm	350 rpm
Approximate monthly kWhs @ 12 mph	124 kWh	188 kWh	158 kWh	192 kWh
Rotor Diameter	8.4 feet	8.2 feet	9.0 feet	11.8 feet
Maximum design wind speed	145 mph	120 mph	120 mph	>100 mph
Turbine Weight	154 lb.	75 lb.	65 lb.	250 lb.
Direct Grid-Tie	no	no	yes	yes

SOURCES: Bergey Windpower, Southwest Windpower, Solar Wind Works, Abundant Renewable Energy, *Home Power* Magazine

Towers: Holding Your Turbine Up in the Breeze

Note: *Wind tower grounding is covered on page 215.*

Once you have a pretty good idea what size and type of wind turbine will fit your needs, you'll have to figure out how you're going to hold it up in the path of the wind. There are four basic types of towers used by most homeowners: guyed-pipe, guyed-lattice, free-standing lattice, and tubular monopole towers.

Pipe towers are the cheapest, easiest to set up, and probably the most widely-used. Made from sections of standard, off-the-shelf galvanized steel tubing, they are sleek, slim, and as inconspicuous as a tower can be—which isn't very. They are hinged at the base and then erected with the turbine already installed, blades and all. The major drawback to a pipe tower is that you cannot climb it for periodic inspections; it must be lowered.

Guyed-lattice towers are like ham radio towers. They are 3-sided and of uniform dimension from top to bottom—mine is around 18 inches per side—and, like pipe towers, must be supported by a series of guy wires. They can be assembled either vertically by sections, or on the ground on a hinged base and tilted-up into place.

Free-standing lattice towers are broad at the base and taper toward the top, much in the same

Proven WT2500 on an 84-foot guyed-pipe tower in Nevada. *Photo courtesy of Solar Wind Works.*

elegant way as the Eiffel Tower. Though more expensive (and showy) than pipe or guyed-lattice towers, you won't have to worry about clothes-lining yourself on a guy wire whenever you walk near it. Like the guyed-lattice towers, free-standing towers can be built in place, or assembled flat and raised to their vertical position.

A fourth type of tower, currently being offered by Bergey for their 10 kW turbine, is the free-standing tubular monopole tower. These are like the solid, tapered steel towers you see holding communication equipment, or lights high in the air above highway exit ramps. They're expensive and require a crane to erect, but they're solid and good-looking, and they take up very little ground space.

Most turbine manufacturers offer tower kits sized for each of their turbines, and those who don't sell the kits will make recommendations on which towers will work best with a particular turbine. Listen to these folks—they know what it takes to hold their machines up in the wind. The lateral thrust put on a turbine in a high wind is mind boggling, and nothing you want to experiment with. You wouldn't put a V-8 in go-cart, would you? Same difference.

A free-standing lattice tower does not need guy wires. *Photo courtesy of Bergey Windpower.*

But even with good engineering, your tower and its foundation, like any structure, may be subject to regulation by your local building department. This means that you will have to comply with whatever codes are in place, since, unlike an un-permitted workshop tucked inconspicuously away in the trees, it's a bit difficult to erect a tower without anyone noticing.

In any event, there's no substitute for sound engineering, so unless you're an engineer and rigger by trade, you should seek professional assistance to ensure that all goes smoothly.

I personally prefer a lattice

tower for the simple reason that it's easy to climb when I want to conduct periodic inspections. I can also climb the tower (using proper safety equipment, of course) to wipe ice from the propeller blades after a freezing rain, rather than waiting a day or more for the sun to come out and melt the ice away.

Our lattice tower is a 50-foot 1950s affair, that stood un-guyed for 40 years with a radio antenna mounted on top before I took it apart and hauled it up the mountain. It now rests in a 3-cubic yard block of concrete, set in solid bedrock. It's nine ¼-inch wire-rope guys are likewise anchored in bedrock. I'm pretty sure it could ride out a hurricane with a Volkswagen parked on top, though I doubt I'll ever find out. (As much as I'd like to tell you how I raised it, you're probably better off not knowing.)

Looking Deeper

In this chapter I have tried to address every aspect of wind power that I deemed necessary for you to consider before deciding what type of turbine to buy, and where—and on what—to mount it. At the very least, it should help steer you in the right direction. I know that LaVonne and I came to our decision with far less information than is presented here.

At the same time, I realize that the information contained here is, at best, a shallow scratch on the skin of a very deep body of knowledge. There are volumes of literature available on the subject of wind power. Hundreds of incredibly brainy people have devoted their lives to the study of harnessing the wind for the purpose of providing reliable, renewable, non-polluting energy. Yet, for all of that, it remains an elusive, empirical science. No matter how efficient an airfoil one person designs, someone else will always find a way to make it better. The same goes for turbines, and the associated electronics. There will forever be room for improvement, and that's the way it should be.

The wind is, after all, just so much air.

Personal Power Companies

Homeowners	Ron & Gretchen
# of Occupants	2 adults
Location	Golden, Colorado
Home Size	2,700 square feet on 2 levels
Home Heating	Solar-heated in-floor hydronic, plus 2 secondary-combustion wood stoves
Water Heating	Solar-heated water with on-demand electric water heater for backup
Grid-Tied?	Yes, at the option of the owners
PV Array	7,200 watts; roof mounted (see book cover)
Charge Controller(s)	2 OutBack MX60s
Inverter(s)	2 Xantrex SW5548s
Batteries	32 Trojan L-16, flooded lead-acid; 1,560 amp hours @ 48 volts
Wind Turbine /Tower	None
Solar Hot Water	Evacuated tubes and flat panel collectors; 2 insulated concrete tanks, totaling 10,000 gallons, are buried below house
Backup Generator	None
Comments	Ron & Gretchen's home is so intensely solar because it was built around the CU 2002 Solar Decathlon House. The flat-panel solar collectors are recent additions, as are the gargantuan hot water storage tanks, but the key components of the PV system are original. This house produces far more energy than Ron and Gretchen could ever use, so the excess will be sold back to the utility. To make the best of the electrical bounty, they are looking for a street-legal electric car to plug into the system.

## chapter 7

Micro-Hydro Power

developing a meaningful relationship with gravity

When I was a little boy I was fascinated by the giant water wheel my grandfather had set up in the irrigation ditch that ran next to my grandmother's garden. Round and round the wheel turned in the swift current, catching water in soup cans fastened to the paddles. Just past the high point, the cans emptied into a trough that routed the water into all the little channels that ran amongst my grandmother's carrots, radishes and peas. I would sit for hours in the branches of the apple tree that hung over the ditch, eating green apples until I was sick, and marveling at the industriousness of the ceaselessly turning paddlewheel.

It was my first encounter with hydro power.

My second encounter came years later, when I spent the summer with my brother on his cattle ranch in Costa Rica. There were two villages on the ranch, separated by 3 miles of jungle and pasture, and several hundred feet of elevation. The high village had no electric power, other than that supplied by a giant diesel generator. Every night for 3 or 4 hours the generator would run so people could have electric lights for the time between sunset and bedtime, to read or sew, or do whatever they did under artificial light. (The Ortiz family had the only TV in the village, and when the generator was running the local children would swarm to the big picture window in front of their house like so many moths to a yard light.)

> ## Dam Facts
>
> The turbine intake pipes (penstocks) for China's Three Gorges Dam are over 40 feet in diameter, big enough to contain an average two-story house.
>
> The Reisseck hydroelectric plant in Austria boasts a head of 5,800 feet, more than one mile. The water pressure at the turbines is over 2,500 psi.

By a serendipitous circumstance of geography, the second village had no need of a big, noisy diesel generator. Situated at the bottom of a hill 600 vertical feet below a fast moving stream, it had all the power it needed to supply the modest needs of a few primitive homes and a small commercial sawmill. A portion of the water from the stream above was diverted through a 12-inch pipe and sent shooting at incredible speed and pressure through a powerful turbine, before emptying into the large river just below the sawmill.

I never knew all the particulars of that system—the output of the turbine's generator, the speed of the water, or how much water pressure developed in the pipe—since I didn't give a whit about renewable energy at the time. I did know that the power was always there

> ## Sunlight
>
> Like solar and wind energy, the ultimate source of hydroelectric power is sunlight—as opposed to gravity, as you might be thinking—since it was solar radiation that got the water (by evaporation and condensation) to the top of hill in the first place.

for the taking, and it was free, though is was not as spellbinding as my grandmother's paddlewheel.

Micro-hydro power is a dream come true, for those lucky enough to be able to use it. As marvelous as solar and wind technology may be, you are always at the whim of the weather. With hydro power, on the other hand, the energy you get when water flows from a high place to a lower place is fairly constant and, as long as there is water flowing, it takes the guesswork out your daily energy equations.

It's a little like the difference between day-trading in volatile stocks, or putting your money into high-yield bonds: with the former you may have a killer day (a steady 30-mph wind, with the sun beating down from an azure sky), but you could just as easily go bust (foggy and calm); with the latter (flowing water) you know exactly what you're getting, day in and day out.

The question is, do you have the right currency to get into the game?

Sizing-up the Possibilities

If you want to create significant amounts of electricity, running water won't do you much good unless you are also able to use gravity to develop pressure. That's because the best way to create water pressure is to pour water on top of water. Commercial hydroelectric operations accomplish this by building giant dams across mighty rivers, then running the water through huge pipes (penstocks) into turbines located below the dams. The sheer weight of all the water above and within the penstocks creates the pressure (.434 psi per vertical foot of water) needed to spin the giant turbines.

Micro-hydro electric systems differ from mega-hydro systems in that you don't have to create a monster lake out of your scenic, bubbling brook to make good use of the power. Instead, you simply collect a portion of your stream's water upstream, then route it downhill through a penstock (a pipe, generally in the range of 2 to 4 inches in diameter), where it naturally develops pressure. Once the water passes through the turbine it returns to the stream, unchanged except for the loss of a little kinetic energy.

In order to use micro-hydro power, then, you need to have a requisite volume of water running down a fairly good slope. The greater the slope, the more power you'll be able to generate, since you will be able to achieve more "head" (the vertical drop from the water intake to the turbine) with a shorter penstock, and lose less power due to friction within the pipe.

Assuming that you have the volume, and the head, all that remains to make this a perfect setting for a micro-hydro turbine-generator would be to have your house close to the stream, so there is little loss of power in the electrical lines running between the turbine and the batteries used to store the power.

But most of us are not quite so lucky. You're not likely to have a house right next to the stream, and the stream might not have the volume—or the drop—you'd like it to. So maybe you won't be able to consume free energy with utter abandon until the end of time. But you still might be able to get enough use out of your stream to augment a solar and/or wind system, and increase your comfort level proportionately.

Basic Components of a Micro-Hydro System

Penstock
Turbine-Generator
Regulator & Diversion Load
Batteries & Inverter

The Components in Context

■ The Penstock

The penstock, or piping used to carry water to the turbine, can be made from a number of materials, including steel, plastic, or most commonly, PVC (polyvinyl chloride). It should run as straight as possible, with few sharp bends.

You will need a strainer over the water intake to keep debris from entering the penstock and plugging the nozzles supplying water to the turbine. In cold-weather climates the pipe should either be buried or insulated to protect it from freezing.

■ The Turbine-Generator

The heart of any micro-hydro system is the turbine-generator. Similar to the turbine for a wind generating system, a micro-hydro turbine uses high-pressure jets of water to turn a propeller, also called a runner. The runner is attached to a shaft running through the generator (or alter-

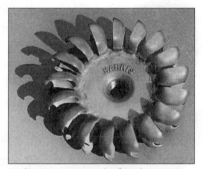

A pelton-type runner made of cast bronze.

nator) which spins a powerful magnetic rotor. The rotor induces an alternating electric current within the windings of the stator that surrounds it, which is then (at some point) converted into DC current, so it can be stored in a battery bank. As with a wind generator, higher rpm's produce more amperage, so you can think

A 4-nozzle P.M. (permanent magnet) generator-equipped turbine. *Photos courtesy of Harris Hydroelectric.*

of volume and head of water as the equivalent of wind speed.

Most high-head micro-hydro systems employ a Pelton wheel for the turbine runner, which is a propeller with bucket-shaped depressions to catch the water. One to four high-pressure nozzles supply water to the runner. The number and size of nozzles used is determined by the volume of water coming through the penstock.

How much head do you need for this type of turbine? Depending on the volume of water making its way to the runner, you could begin producing usable power with as little as 20 feet, though more is certainly better.

I should also mention that, since the generator attached to the turbine will probably not be designed to run underwater, you should plan to locate your turbine-generator along the side of the stream, above the flood plain.

■ The Regulator (Charge Controller) and Diversion Load

As with solar and wind power, a regulator (or charge controller) is needed to keep the batteries from becoming overcharged. A solar charge controller won't work, however, since these types of controllers simply disconnect from the power source, once the batteries reach full capacity. This would leave the turbine to spin dangerously fast, causing a lot of expensive problems.

You will need a way to safely dump any excess amperage, without leaving an open circuit. A diversion load, in other words. As with wind generators, the best way to bleed off excess power is to use it to produce heat, such as with a space heater or a water heater. In some systems the excess power is shunted to the heat sink before it reaches the batteries; in other systems the turbine's generator is hooked directly to the batteries, and any excess power is directed to the heat sink from the backside of the batteries through a diversion-load controller (a good solar charge controller will work for this purpose, provided it is properly programmed; *see Chapter 8 on Charge Controllers*). Again, the manufacturer or system designer will be able offer various options, based on the specifics of your particular system.

◼ Batteries and the Inverter

Once the electrical current passes the charge controller (and/or is shunted to the diversion load), your micro-hydro system will require no other special components. The current can be run into the same battery bank you use for your solar and wind, and can likewise be run through the same inverter. The only caveat to this would be if your hydro-generator was located a considerable distance from your house (as it might well be). In that case, the size and cost of the wire needed to carry the low-voltage current from the turbine to the rest of the system could become unwieldy *(see the line loss tables in the appendix for determining wire size for various amperages and distances)*. If that's your situation, you might want to consider setting up a complete system (batteries, inverter and all) close to the stream, then running 120VAC to the house, through smaller wires. Of course, this arrangement comes with its own set of problems, not the least of which is the extra cost involved in doubling-up on expensive com-

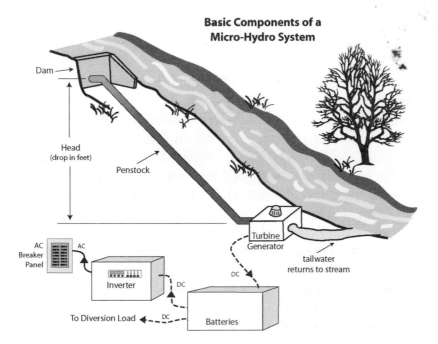

Basic Components of a Micro-Hydro System

ponents, in the event that you are also using solar and wind generating sources. Which way should you go? Unless you're a whiz at this sort of thing, this one is best left up to a qualified system designer.

Getting Down to Particulars

If you suspect you might have a viable micro-hydro site, the best thing you can do is to take a few measurements and then plug the numbers into the formula given at the end of this section. It will give you a rough idea of the amount of energy your site will produce. In particular, you will need to know the five variables for micro-hydro power:

1 **Length of pipe** from the water source to the turbine
2 **Feet of vertical drop** from the water source to the turbine
3 **Flow rate** of your stream (the portion delivered to the turbine)
4 **Length of the wire** (from the generator to the batteries)
5 **System voltage**

■ **Rough Calculations of Power (using variables 1, 2 and 3)**

The length of pipe and the feet of drop may take a little trial and error measuring. As I said before, the greater the drop the better. Even though there will be more resistance (friction) in a longer penstock, the extra pressure will more than make up for it.

Start with a likely spot upstream for collecting water in the penstock, then determine how much higher it is than the place where you wish to locate your turbine. This will give you your gross head. A quick and dirty way to do this is to start from your turbine site by sighting down a 4-foot level at a given height above the stream—you'll need an assistant to confirm that the level really is level—and marking the spot where your eye meets the water, upstream. Then go to that spot and repeat the process. You can also use an altimeter (such as the ones found on pricey watches and GPS units) to get a fairly good idea, providing the atmospheric pressure does not change appreciably from one measurement to another. Reading the contour lines on a small-scale topological map will also yield a rough idea of how much head you'll have.

Once you know how long your penstock is going to be, and how much drop you'll have, you need to know how much water you can collect in the penstock. Try building a small, temporary collection dam on the side of the stream and directing the water through a large pipe. Then time how long it takes to fill a 5-gallon bucket or some other large container. A couple of simple calculations will yield the gallons per minute. Now, since our formula requires cubic feet per second instead of gallons per minute, we need to make one last conversion before applying the formula.

Let's say, for example, you've got a flow of 100 gallons per minute. To convert it to cubic feet per second, divide 100 gallons per minute (gpm) by 448.8, the number of gpm's it takes to equal 1 cubic foot per second (cfs): 100 gpm ÷ 448.8 = .223 cfs.

Finally, we're ready to run the formula. For the sake of our example, let's say that we have determined that we will have 50 feet of gross head, and we assume 55 percent efficiency. Our formula goes as follows:

Gross Head x Flow x System Efficiency x C = Power
Example: 50 ft. x .223 cfs x 0.55 x 0.085 = .52 kW (or 520 watts)

We see that our little system will produce somewhere on the order of 520 watts on a continual basis. This is equal to nearly 12.5 kWh's per day, which is plenty for many homes without wasteful appliances, where the residents are mindful of their energy usage. LaVonne and I live quite comfortably, in fact, with half that amount from our wind and solar systems, and many of our friends get by on much less.

If you play with this formula a little, you will make an interesting discovery—namely that 100 feet of gross head delivered at 50 gallons per minute is equal to 50 feet of head delivered at 100 gallons per minute. An exceptional flow, then, can compensate for low head.

This formula really gives just a ballpark number. You should use it to determine if it's worth the trouble to go to the next step. You will

- **Gross head** is the actual distance of drop from the intake to the turbine (in feet or meters), not taking into account the friction developed in the penstock.

- **Flow** is measured in cubic feet per second (cfs), or cubic meters per second (cm/s).

- **System efficiency** will be between 40% and 70%. *It is affected by the length and condition of the penstock, the efficiency of the turbine, the gauge and length of the wires running from the turbine to the batteries, etc. If you're planning on buying a state-of-the-art turbine, having a relatively short distance from the turbine to the batteries, and do not have an inordinately lengthy penstock, start with 55%.*

- **C** (the constant) is 0.085 when using feet; 9.81 when using meters.

notice, for instance, that the length of the penstock is not entered into the equation directly; instead it is one of the factors you take into account when figuring the efficiency factor.

■ **Refining the Numbers (using variables 4 and 5)**

By this point you should know if you have a feasible micro-hydro site. The next step is to contact someone who can refine your numbers and make specific suggestions for the system components. Now you will need the last two variables listed earlier (4 and 5): the length of the wire running from the generator to the batteries, and the system voltage.

This should be reasonably straightforward, especially if you already have a solar and/or wind system in place. If not, pick a likely spot for your batteries (inside a heated building, if possible) and measure the distance. If it's greater than 50 feet, you should consider setting up a 48-volt system to minimize line loss.

Once you have all these facts in hand, you'll need to find someone who can make sense of them. You could hire the services of an engineer, but you shouldn't have to. Most manufacturers of turbines will be happy to run the numbers for you. Or, for an analysis from a company that sells more than one type of system, give Real Goods a call. In any event, it never hurts to have more than one opinion (or one bid).

If your numbers bear fruit and the cost of the system is not enough to make you blanch, then there is one more important step you need to take before ordering your components: you will have to talk to a county agent about the legal implications of your proposed plans. If the amount of water you intend to divert from the stream is but a fraction of the overall flow, then you will probably not encounter much resistance. But you still have to ask; water—and the natural habitats it supports—is a very touchy subject these days, and not one those issues where forgiveness comes easier than permission.

Finding the Exact Efficiency of Your System

If you do install an actual system, you can see how close you were to guessing the efficiency factor by running the formula backwards:

kW ÷ (Gross Head x Flow x Constant) = True Efficiency Factor

MICK'S MUSINGS

Solar panels need sun to excite them; wind turbines need moving air; hydro turbines need flowing water. Dogs only need cats.

The Jack Rabbit

Do you have a high-flow, low-head stream? Or you just aren't ready to sink a ton of money into a complex micro-hydro system? The Jack Rabbit low-head turbine from Jack Rabbit Marine Energy Systems may be the answer. Though it's only rated at 100 watts, that's still over 2 kWh's per day which,

depending on where you live, is equal to the average output of 400 to 600 watts of solar modules. The best part is, it doesn't require fancy engineering or hours of labor to make it work. You simply make a constriction in the stream (with logs or rocks, or whatever) to speed up the water—so it flows at least 9 mph—and mount the Jack Rabbit facing upstream.

photo courtesy of Real Goods

Power in the rough—a micro-hydro enthusiast's dream. *(spillway on the Big Thompson river, near Estes Park, Colorado)*

## chapter 8

Charge Controllers
processing your batteries' diet

The first time I ever held a charge controller in my hands, I was far from impressed. What could be so special about a small metal box? It wasn't until I set it aside and began to read the instruction booklet that I began to appreciate just what a marvel of electrical engineering a charge controller is.

The primary purpose of a charge controller is to charge the batteries, without overcharging them. Different controllers have different ways of achieving this objective. Some work better than others.

A charge controller will also disconnect the battery from the solar array after dark to keep current from flowing out of the batteries and back into the modules as they sit idle.

OutBack's MX60 charge controller. *Photo courtesy of OutBack.*

There are lots of inexpensive (under $50) charge controllers on the market that offer this bare-bones type of performance for small, unsophisticated systems, but for units capable of battery equalization or multi-step charging, be prepared to spend a bit more money. The more sophistication you can add to the processes of charging and equalizing, the better off you'll be.

Battery Charging

We put a Trace (Xantrex) C40 charge controller, with a C40DVM digital meter, in the frame cabin where we lived while building the log house. It charges the batteries in three distinct stages: first, it allows the full charge from the PV array to reach the batteries, until a preset voltage limit is reached. This period of unrestricted charging is called the bulk stage.

Once the bulk voltage setting is reached, the controller backs off the amount of current sent to the batteries, in order to hold the voltage at the bulk setting for a cumulative period of one hour (this is the absorption stage).

After that, the controller enters the float stage, where the voltage is allowed to drop to a lower (preset) voltage, where it will be maintained until the sun sets, or the (AC) loads exceed the DC input. The bulk and float voltage settings are determined by the installer, and are set to voltages most practical for the specific application and the types of batteries used.

Why the complexity? Simply put, the various stages are needed to allow the batteries to "soak-up" a charge. If you use a multimeter to read the voltage of a battery as it's being quickly charged, and then disconnect the charger and take another reading, you will notice a significant voltage drop. If you let the battery sit for several minutes and take still another reading, you will see that the voltage has dropped even further. It may seem that the battery is mysteriously losing its charge, but it isn't; it's merely dispersing the charge throughout the cells. By charging the batteries in stages, the charge controller ensures that the batteries reach an actual full charge rather than an *apparent* full charge.

Xantrex charge controller.
Photo courtesy of Xantrex.

Equalization of Batteries

Equalization is the second important function of a charge controller for flooded, lead-acid batteries (not sealed). Sulfates can build up on the batteries' plates over time, and affect their performance. If the sulfates crystallize on certain areas of the plates, those areas are no longer able to function. By bringing the batteries to a very high state of charge, most of the damaging sulfates dissolve back into solution, increasing the batteries' storage capacity.

Blue Sky's 3024i charge controller with remote sensor. *Photo courtesy of Blue Sky.*

The C40 equalizes the batteries by holding them at one volt above the bulk setting (2 volts for a 24-volt system, or 4 volts for a 48-volt system) for a cumulative period of 2 hours. The charge controller may be set to equalize automatically every 30 days, but that's probably too often for a battery bank that is never deeply discharged. Besides, I prefer to initiate the process manually. That way, I can pick a sunny, windy day when I know the batteries will equalize quickly.

A Charge Controller for the Wind Turbine

Since our wind turbine sends 3-phase AC to the house, it came with its own charge controller, one that uses a trio of rectifiers to change the AC into DC. Because it also has leads for 40 amps of DC input from the solar array, we used this charge controller for both sun and wind at the new house.

Initially, I was quite impressed with it. Its big LED display offered a wealth of information, including separate readings for wind and solar amps, battery voltage, and a "volts per cell" feature (per each 2-volt battery cell, that is) that was internally averaged over time to provide a fairly accurate "fuel gauge" for the batteries.

With this type of charge controller, battery charging is regulated by a relay that sends excess amperage to a heat sink, once a preset volts-per-cell limit is reached. Equalization is achieved by setting the volts-per-cell dial to a higher (equalizing) limit, and leaving it there for a day or two, or until all the individual battery cells test the same with a hydrometer. It was a tedious process that made me long for the sophisticated simplicity of a C40 solar charge controller.

After four months of living in the new house, we added 480 watts of solar modules. Because the combination wind/solar charge controller was not rated to handle the additional solar load, we bought a Trace (Xantrex) C60 charge controller and diverted the entire solar input through it. We were then able to regulate battery charging and equalization with the C60, while using the wind charge controller's

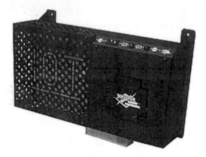

regulating features only to ensure the batteries don't become overcharged in high winds. Each controller senses the batteries' state of charge and adjusts its output accordingly. We have had no problems using the two controllers together.

Whisper controllers include a diversion load to ensure safe operation of the wind turbine when the batteries are fully charged. *Photo courtesy of Southwest Windpower.*

Maximum Power Point Tracking (MPPT)

Blue Sky Energy, OutBack Power Systems and others have charge controllers that maximize the charge by converting excess array voltage into usable amperage (power point tracking). This can be particularly helpful during the winter months, when the modules are cold and therefore operating at higher voltages. Additionally, look for models that can use 48-volt current from the array to power a 24-volt system. This can be very helpful for long wire runs if you're locked into a 24-volt system but your array has grown too large for the skimpy wire you have running to it.

■ Squeezing the Last Watt From Your Array

Solar modules are designed to operate at higher voltages than the batteries they're asked to charge. There are two main reasons for this. Since voltage always flows from a higher potential to a lower one, the modules need to operate at a high enough voltage to charge the batteries in both low light conditions (when the array voltage drops), and when the batteries reach a high state of charge. Since lead-acid batteries (in a nominal 12-volt system) often reach potentials as high as 15.5 volts during equalization, the array's rated voltage must be even higher. Typically, this will be in the range of 16.5 to 18 volts. During the bulk charging stage (the stage most charge controllers work in, most of the time) a typical charge controller will simply hook the array directly to the batteries. The batteries, of course, don't have any use for all that extra voltage, so they pull the array voltage down to a comfortable level.

"So what's the big deal?" I can hear you say. "If, as you've led us to believe, watts = volts x amps, why doesn't the amperage simply go up, as the voltage drops?" Good question. The answer is it can—to a point. That point is the amperage the module was designed to produce. Think of it as a brick wall, 'cuz you can't get past it. At least not with a conventional charge controller.

Let's take a typical module: a Kyocera KC120. It's rated at 120 watts, when it's producing 7.10 amps at 16.9 volts (7.10 x 16.9 = 119.99). How does this play out in the real world? Well, if your batteries are at 12.5 volts and they're drawing all 7.10 amps, the module is only producing 88.75 watts, or about 74 percent of its rated output. As the batteries reach a higher state of charge the percentage will go up, but there is always a significant percentage of power that is lost to heat.

An MPPT (Maximum Power Point Tracking) charge controller gets around this problem by using a DC to DC converter. It takes whatever voltage is optimal, and converts it to the voltage the batteries are happy with. In the process, it uses the extra voltage to produce usable amperage in excess of what the modules were designed to produce. The trick is that the modules don't know this. They just think the batteries are finally getting their act together.

Power point tracking works best, then, when there is a large disparity between the PV module voltage, and the battery voltage. This most often occurs when the batteries are partly discharged, under a considerable load, or when the modules are cold. As the batteries reach a higher state of charge, loads are decreased, or sunlight heats up the modules, the extra power gained from MPPT diminishes.

This leaves us with the question: what is the power point that's being tracked? To find the answer, you'll have to imagine playing a game—a board game. Unlike checkers or Monopoly, however, the board is completely black, except for faint calibrations on the left side and bottom. The left-side marks represent amperage; the marks across the bottom are for voltage.

The board is plugged into your array. You have been assigned the role of Charge Controller. You firmly grasp your stylus and flip the switch. A thin bright line appears across the board; a power curve. It's sunny out and nearly midday, so the line runs almost straight from near the top of the board to the right, all the way over to where it meets the array's rated voltage, around 17 volts. From there it curves down and meets the edge of the board at the open circuit mark of 21 volts on the bottom right side.

You place your stylus on a point on the line. Instantly the board lights up below and to the left of your stylus. The lighted area represents wattage, and it's your job to light up as much area as possible.

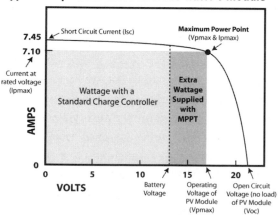

Typical Amps-Volts Curve for a 120-watt PV module

To make it easier for you, a small display appears to show you how much area of the board is lit up. You notice that, as you move your stylus up and down the curve, the area changes. If you go too low on the curve your voltage increases, but your amps are so low that the total power begins to diminish. Move up the curve, past where it straightens out, and you find that you've got lots of amperage, but too little voltage to maximize the wattage.

Finally you find the optimal spot. Pat yourself on the back; you've just tracked the power point.

If you do this all day with a constantly changing power curve—as the clouds come and go, as the sun moves across the sky and the array heats up and cools off—you'll know just what it feels like to be an MPPT charge controller.

Of course, if you wanted to be a regular charge controller, you'd just touch your stylus on the power curve directly above the battery voltage and call it good. Easy, but not much fun.

The MX60 Charge Controller and the Cheap Freezer

Not long ago LaVonne and I did the unthinkable: we bought a cheap little 5.5 cubic foot freezer from a discount store, wondering if it'd be a ravenous energy pig. We were pleasantly surprised. At 400 watt hours per day (according to our Watts Up? meter) in our 60 degree garage, it's nearly as watt-conscious as a high-efficiency model selling for seven times the price. But 400 watts is 400 watts, whether it's warm or cold, sunny or socked-in. It's more energy than it would take to run a 15-watt compact fluorescent day and night, or to pump 50 extra gallons of water per day from our 540-foot well with a 240-volt, 1.5 hp well pump.

What were we thinking?

Other than the circulating pumps for the hot water boiler—an energy demand we allowed for when originally setting up the system—there were no large compulsory energy drains. So, efficient or not, we knew our little freezer was going to take a painful bite out of our energy budget.

At first it wasn't too bad. Our customarily stalwart system handled the freezer well enough through the fall months, but once gloomy winter weather set in I found myself running the gas generator more often. My first temptation was to buy another 200 or 300 watts of solar modules to charge the batteries faster whenever the sun came out of hiding, but LaVonne—who views money in far more tangible terms than I—suggested we look for another, less expensive, way to increase our daily harvest of precious watts.

That's when I looked at OutBack Power Systems' MX60 charge controller. If its Maximum Power Point Tracking (MPPT) technology really worked, it would save me the time and expense of expanding our array. The MX60's circuitry would also allow me to rewire the array for 48 volts—to minimize any line loss from the array to the controller—while still keeping the rest of the system at 24 volts.

The price tag on the MX60 was over three times what we paid for our old Trace C-60, but if it gave us an extra 75 to 100 watts, on average, throughout the sunny part of the day, it would be well worth the extra money.

Happily, the MX60 did the job and then some. Where before our array was maxed-out at 33 amps, I was now regularly seeing the amps in the 37 to 42 range, or even higher for short periods. The difference is especially noticeable on cold days when the sun shines through high clouds, or when direct sunlight hits the array after being obscured by heavy clouds. Once I wired the array for 48 volts the output was even stronger—about 5 percent, overall—than at 24 volts.

Thanks to the MX60 I no longer snarl at the freezer when I walk past it and hear the compressor running.

Other Uses for Charge Controllers

Some charge controllers have been designed to do more than regulate battery charging and equalization. Specifically, they may be used as either diversion load controllers, or as DC load controllers (though not at the same time as they are being used as charge controllers). Most of us have little need for either of these extra functions, but a short explanation of each may save you a moment or two of bafflement as you read the manufacturers' operating manuals.

■ Diversion Load Control

What's the purpose of a diversion load controller? Well, let's say you have an old-style DC wind generator (or a micro-hydro turbine) wired directly to the batteries. If the wind blows hard enough and long enough, the batteries could become severely overcharged and possibly ruined.

Big deal, you say. Why not just run the wind generator through the charge controller, and let it deal with the excess charge the same way it does with a solar array? The reason you can't do this is because a solar charge controller will simply disconnect the source from the batteries when the voltage gets too high; do this with a wind generator and the propeller will spin far faster than it was designed to spin.

By placing the charge controller (turned diversion load controller) on the other side of the batteries from the wind generator, and using it to divert excess current to a heat sink, the wind generator remains attached to the batteries at all times without overcharging them, since any excess sent to the battery bank will be drawn off the other side.

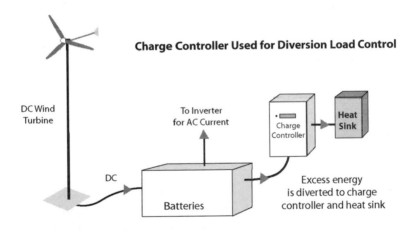

Charge Controller Used for Diversion Load Control

DC Wind Turbine

To Inverter for AC Current

Charge Controller

Heat Sink

DC

Batteries

Excess energy is diverted to charge controller and heat sink

■ DC Load Control

Let's say you have a DC refrigerator that runs directly from the batteries. What happens when the batteries get too low to operate the compressor motor? It could damage the motor, and would certainly damage or destroy the batteries. If it were an AC appliance, the inverter would simply disconnect itself from the load, until the batteries were recharged sufficiently to once again supply ample current. But a DC load doesn't go through the inverter, and so is afforded no such protection. Unless, of course, some other regulating component is installed between the DC load and the batteries. That's the purpose of a DC load controller: when the battery voltage falls below a preset level, the load controller disconnects the load until the batteries again reach a safe level of charge. If the low-voltage condition persists for an extended period of time—as could happen if your array was buried in snow and you were away from the house for a few days—it may cost you a lot of food (and perhaps subject you to a memorable olfactory experience), but it's still a ton of money cheaper than a new high-efficiency refrigerator and a dozen or so batteries.

Charge Controller Used for DC Load Control

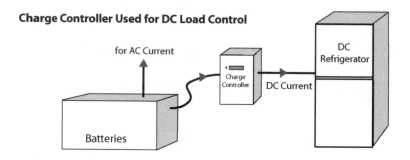

Sizing for the Future

As with everything else, buy a charge controller that will be big enough to handle the extra amperage, should you decide to add more solar modules at a later date. Or design your solar setup in such a way that it can be split into two arrays, operating through a pair of charge controllers.

Unlikely Charge Controllers

What do a refrigerator's ice maker, an AC pond pump, and a charge controller have in common? They can all be used to regulate how much charge goes into your batteries.

What am I talking about? Well, when I finally broke down and told LaVonne we could get a "real" refrigerator—one that runs on electricity instead of propane—I never realized how useful it would be as a tool to regulate the batteries. The off-the-shelf Kenmore fridge we bought was rated at 381 kWh per year by the strict criteria they use to make such determinations. But we quickly learned, with the help of our Watts Up meter, that with the ice maker turned off, the fridge used only around 275 kWh per year, while it would use over 400 kWh per year if the ice maker ran 24/7, which it never would, since no one could ever use that much ice. Hence, on days when the solar array is cranking out wattage faster than we can use it, we make ice all day. On cloudy days we shut off the ice maker, use stored ice, and let the fridge coast at the lower energy usage.

The same goes for the AC pond pump we installed in our new pond. Since it draws around 40 watts it's hardly anything we'd want to run all the time, so the fountain of water it shoots into the air is a pleasant sunny-day extravagance. I consider it a reward for diligently watching the watts when it really counts.

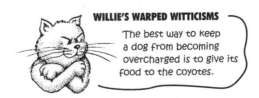

WILLIE'S WARPED WITTICISMS

The best way to keep a dog from becoming overcharged is to give its food to the coyotes.

Personal Power Companies

Homeowners	Michael & Dianne
# of Occupants	2 adults
Location	Cashiers, North Carolina
Home Size	2,700 square feet on 2 levels
Home Heating	Jøtul wood stove; in-floor hydronic with propane boiler
Water Heating	Propane boiler (with future solar)
Grid-Tied?	No
PV Array	3,840 watts; roof mounted
Charge Controller(s)	Tristart TS-60
Inverter(s)	Trace (Xantrex) SW5548 sine wave
Batteries	2 Deka M100-17 flooded lead-acid; 1,600 amp hours @ 48 volts
Wind Turbine /Tower	None
Solar Hot Water	Piping roughed in but system not installed
Backup Generator	None
Comments	This is a well-planned, high-production system, designed in large part by Michael and Dianne with help from North Carolina State University's Solar Center. Note that this is a true stand-alone system, with no backup generator. Domestic water is pumped into a cistern from 500 feet with a Dankoff 48-volt DC low-volume pump.

## chapter 9

Batteries

care and feeding of the beasts that hold your sunshine

Wind turbines whir, inverters hum, but batteries just grumble. And why shouldn't they? Compared to the workout the batteries are subjected to every day, the other components in your PV/wind system have it pretty easy. Sure, the wind turbine is stuck up on a tower, being beaten around by stray updrafts and errant cross currents all day and night, but it's a feisty machine—it loves to be in the thick of the action. The batteries have to stay forever locked up in a dark box, with no idea what's going to happen next. They alternate between being overstuffed with wattage to wondering where their next meal is coming from. They're in a constant state of indigestion.

Believe me—if anything in your system deserves a little pampering, it's the batteries.

How Many Batteries?

See *Chapter 4: Sizing The System*, and the battery sizing worksheet on page 266.

Trojan T-105 and L-16H batteries. *Photo courtesy of Trojan Battery Company.*

> **Be Careful! Batteries are Dangerous!** Flooded lead-acid batteries vent hydrogen gas, which is highly flammable. As if that weren't enough, they are filled with a caustic brew known as sulfuric acid. Always vent batteries to the outside, and don't let them anywhere near sparks or flames. And never, ever, allow a conductor (such as a wrench) to come in contact with terminals of opposite polarity. It could cause an explosion!

The Right Batteries For The Job

What is the best thing you can do for your batteries? Never ask them to do more than they were designed to do. Only true, deep-cycle batteries will successfully do the job you have in mind. Because they have thicker plates with less surface area than automotive batteries, they are made to be charged slowly, and don't mind being discharged deeply.

The first time you reach down and lift a deep-cycle solar battery you will be surprised at how much it weighs, compared to a car battery of comparable size. That's because car batteries are made with thin plates (with lots of surface area) so they can be charged and discharged quickly. The battery in your car (or duty-hardened pickup) is rarely discharged to more than 10 percent of capacity before it's quickly recharged by the alternator. A car battery has it pretty easy compared to what you're going to put your solar batteries through.

A battery that falls somewhere in between a 6-volt deep-cycle battery, and a 12-volt automotive battery, is a 12-volt RV (or marine) deep-cycle battery. You might think that you can use this type of battery and save yourself a few dollars and some confusing terminal connections, but you won't be happy when the batteries wear out in a couple of years. They have thinner plates than the 6-volt deep-cycle battery, and are not made for the rigors of a PV/wind system.

Sealed, maintenance-free batteries (either liquid lead-acid, or gel type) are becoming more popular for PV/wind systems, largely because you never need to check the water level. Couldn't, even if you wanted to. And that's the rub: if they should ever accidentally become overcharged, they would lose part of their water and/or electrolyte through the safety vent, and it couldn't be replaced. Moreover, batteries that have been deeply discharged and left in that condition for a period of time need to be deliberately overcharged to cook the lead sulfate from the plates, but it can't be done (without dire consequences) when the batteries are sealed. So, while they may be less messy, sealed batteries do not allow any remedy, should the plates ever become fouled. You'll be buying new batteries instead of restoring the ones you have.

This is not to say that sealed batteries should not be used, however. In fact, they are the battery of choice for grid-tie systems where the batteries are only called into action during those times when grid power is lost. Batteries in such systems have it pretty easy. They are rarely discharged, and any power that diffuses away is quickly replenished with a maintenance charge from the solar array or wind turbine.

Sealed batteries come in two basic types: Absorbed Glass Mat (AGM) and gel-type. AGM batteries use a liquid electrolyte that is suspended in a fiberglass material that surrounds the lead plates. Because the electrolyte is not free to migrate, it does not become stratified. Makers of AGM batteries include Concorde and MK batteries.

Sun-Xtender® sealed battery by Concorde. *Photo courtesy of Concorde Batteries.*

Gel-type batteries use a true jelled electrolyte, which serves the same purpose as the glass mats in AGM batteries. Because they have more moisture from the get-go, they can afford to lose a little through the built-in safety vents, though they should always be afforded the same care as AGM batteries. Manufacturers of gel-type batteries include MK batteries and Hawker batteries.

Amp hour per amp hour, sealed batteries will set you back more money than flooded lead-acid batteries, but they will last proportionately longer if they are properly maintained—neither overworked or overcharged—in a grid-tied setting.

In off-grid or grid-parallel systems, on the other hand, the batteries are worked hard and will need to be charged and equalized often. For these systems, the only real choice are flooded lead-acid batteries. Probably the most popular battery ever to enter the solar market is the L-16 six-volt battery. Tall (about 17 inches) and heavy, it weighs in at around 120 pounds and holds around 400 amp hours of stored power. These batteries have a life expectancy of 6 to 7 years, but can last over 10 years if properly maintained and never discharged too deeply. Makers of L-16 batteries include Trojan, MK (Deka), and Rolls. Of these, Rolls batteries are the most robust, sport the longest warranty, and come with the highest price tag.

For a lot less money you can set yourself with golf-cart-style batteries. The Trojan T-105, 220 amp-hour battery is probably the most widely used golf-cart-style battery used in small PV systems. With a little more than half the capacity of an L-16, a T-105 battery will cost you about a third as much, but you'll be lucky to see more than five years of service out of it. We installed eight T-105s in the cabin 6 years ago and they're still in excellent condition, owing to the fact that they are rarely discharged to any extent. By contrast, the twenty T-105s we installed in the house five years ago are reaching the end of their useful life. If I can nurse them for another year I'll feel that I've gotten more than my money's worth out of them.

With the skyrocketing popularity of solar and wind renewable energy systems, your battery choices are growing practically by the day. Besides the tried-and-true 6-volt batteries mentioned above, you

---

Spare Batteries?

When some friends up the canyon decided to hook up to the grid, they offered us their bank of twelve L-16 batteries for a mere $100. Even though they were too dissimilar to add to our bank of T-105s, it was too good of a deal to pass up. So for several weeks they sat in the garage, as I ruminated on the possibilities. Finally it came to me: why not install a spare battery bank? By adding an extra breaker on the DC disconnect, I would be able to switch the house loads from one bank to the other, and by adding another breaker below the charge controller, I'd be able to select which bank to charge from the solar array.

Did it work? Even better than I imagined. During sunny, windy stretches I can easily charge up both banks, and we now have an extra 1,200 amp-hours to see us through calm, gloomy spells. Best of all, since we can run the house off of one bank while charging the other, equalization of either bank is a swift, painless process.

---

will find heavy-duty 4-volt, 8-volt, 12-volt and even 2-volt batteries suitable for use in off-grid and grid-tied systems. So look around; somewhere out there is the perfect battery for your application.

■ If You're Really Serious About Batteries....

You'll need a forklift to handle them, but once they're in place you shouldn't have to move HUP batteries for a decade or two. HUP stands for High Utilization Positive, meaning that the gradual corrosion of the positive plates—and the eventual demise of the battery—is greatly retarded by a special, patented process in which Teflon is incorporated into the positive plate material.

Solar-One HUP batteries are all sold in 12-volt units of 6 cells each, pre-assembled in heavy steel cases (euphemistically called trays). Power-wise, the nine different sizes range from 845 to 1,690 amp hours; weight per tray ranges from 642 up to 1,236 pounds. Big batteries, in other words. For system voltages above 12 volts, multiple trays are required.

These are flooded lead-acid batteries, which means they will require the same care and consideration you afford standard T-105 or L-16 batteries. On the upside, with fewer cells there will be fewer caps to fiddle with when it comes time to add water (and since they're all 25 inches high, you won't have to bend over quite so far to check them).

HUP Solar One battery. *Photo courtesy of Northwest Energy Storage.*

To set yourself up with these batteries, you should expect to pay some money; they cost about twice as much as a similarly rated bank of L-16s. But with seven times the life expectancy, they will assuredly pay for themselves over time.

Just don't forget the forklift.

Pampering Your Batteries

Batteries are like draft horses: treat them well, and they will reward you dutifully; treat them badly and they will tire out (or up and die), just when you need them most. And like horses, batteries don't require much to keep them happy; just a warm, dry place to rest, a little water now and then, and the security of knowing they will never go hungry. Nor does either object to being put to work, as long as they're not overworked.

■ Charging and Discharging

As I pointed out in the last chapter, a good charge controller will easily handle the chore of charging your batteries from wind and solar sources, as long as you give it plenty of amperage to work with. But when your batteries become greatly discharged after a few days of heavy loads and/or cloudy weather, you may need to charge them with a fossil-fueled generator. In this case, the charging will probably be done through the inverter, not the charge controller. *(See the next chapter on inverters, for a full discussion of battery charging.)*

How low can you let the batteries get, before you need to drag out the generator? Lead acid batteries can suffer permanent damage if they are ever discharged more than 80 percent of their capacity. This should never happen in a properly-sized PV/wind system. Moreover, a good, well-calibrated inverter with built-in safeguards will shut down the AC loads before allowing the batteries to discharge to such a dangerous degree. *(For large DC loads, a DC load controller should be used as an automatic disconnect. See the previous chapter for more information.)*

To keep your batteries truly healthy (which translates to a long life expectancy), you should never allow them to discharge to below 50 percent of their rated capacity. The trick, of course, is in knowing

---

Battery maintenance takes only 10 minutes every month to check the water level, and look for corrosion on the terminals.

---

when they have reached this level so you can either turn off all the loads connected to the batteries, or use a generator to charge them.

The easiest way, by far, to assess the batteries' state of charge is to install a meter that keeps track of amp hours in versus amp hours out. At a glance, you can tell how many amp hours below full capacity the batteries are, by subtracting the "amp hours from full" from the battery bank's rated capacity. (Our meter, a TriMetric from Bogart Engineering, does the math for us, and displays a "fuel gauge" which shows the state of charge as a percentage.)

I highly recommend a meter, since the alternative is smelly, messy and hard on your clothes at best, and downright dangerous at worst. This alternative is to "weigh" the electrolyte in each battery to determine what percentage of the sulfates are in solution (good) compared to what percent are trapped on the plates (bad). The heavier the electrolyte, the greater the batteries' state of charge.

To accomplish this task, a hydrometer (an apparatus somewhat similar in appearance—though not in function—to a turkey baster) is used to measure the specific gravity of each cell of every battery in the battery bank. (Specific gravity is a ratio that compares the weigh of a given volume of a substance to an equal volume of pure water.) A reading of around 1.172 indicates a 50 percent charge. (This is corrected to a battery temperature of 80 degrees Fahrenheit, so subtract .004 for every 10 degrees less than that.) A higher reading indicates heavier electrolyte with more sulfates in solution, and therefore a more highly charged battery. A lower reading indicates the opposite. A reading of 1.277 indicates the batteries are fully charged, while a reading of 1.098 tells you the batteries are discharged to 20 percent of capacity and in need of immediate charging. When using a hydrometer, beware that it's an inherently messy process. Spilled electrolyte weakens the batteries and wreaks havoc with your clothes. Consequently, it should be used sparingly.

To avoid testing every individual cell, you should be able to get a fair reading by first checking the voltage of each battery with a volt meter, making sure the loads and the charging rates do not vary throughout the procedure. If all the batteries test the same (to within a few hundredths of a volt), test a few random cells with the hydrometer. Significant variation, in either voltage or specific gravity, probably indicates sulfation of the plates, meaning that it's time to equalize.

■ Equalization

Note: *Equalization should only be performed on vented, liquid electrolyte batteries—the kind you add water to, in other words. If you have gel type and/or sealed maintenance-free batteries, you won't need to read this section!*

Equalization is really just a fancy term for the process of overcharging your batteries. The purpose of equalization is to "cook" any sulfates from the plates that may have crystallized there, and also to "stir up" the electrolyte, which tends to become stratified if the batteries go for long amounts of time without being fully charged. When the process is complete, your batteries should all be of an equal charge, hence the term to "equalize."

How often should you equalize your batteries? Expert opinion varies from once a month, to once or twice a year. The Trojan Battery Company (who really *should* know, if anyone does) recommends equalizing only when low specific gravity is detected, or if the specific gravity varies widely (plus or minus .015) from cell to cell, and battery to battery.

Most people, however, will probably not bother to take hydrometer readings at regular intervals. In that case, "better safe than sorry" is the best rule of thumb. Batteries that are brought to a full state of charge often, and rarely (or never) allowed to drop below 50 percent of capacity, should easily be able to go six months without equalization. Batteries that lead a rougher life will need comparatively more attention.

A **hydrometer** measures a battery's state of health; a **volt meter** measures its state of charge. - DOUG PRATT

As mentioned in the last chapter, a good charge controller will initiate and monitor the equalization process. It's all automatic and nothing that requires your attention. There is one detail, however, that I should point out, because it just might save you a lot of hair-pulling later.

The Trace (Xantrex) C-series charge controllers (among others) take the battery voltage 1 volt higher than the bulk voltage setting (2 volts on a 24-volt system, 4 volts on a 48-volt system), and keeps it there for a cumulative period of two hours. So if, for instance, the bulk voltage on a 24-volt system is set at 29.2 volts, the batteries will be brought to 31.2 volts during equalization. There is no problem with this, of course, *unless* the "high battery cut out" setting on the inverter is *below* 31.2 volts. If it is, the inverter will shut down, once the voltage reaches the preset point, and leave you in the dark. You can believe me; I speak from experience.

Hydrometer for measuring the specific gravity of each battery cell.

■ Don't Forget the Water!

Most batteries that land in the recycling heap before their time have simply died of thirst. And most of those batteries come from homes owned by people who have never used renewable energy before. That's unfortunate, because there is really very little work involved in keeping the batteries topped off.

The amount of water your batteries use will depend upon how often they are brought to full charge, how hard you charge them, and how often they are equalized. To begin with, you should check the level at least once a month, and after each equalization. After a few months you will develop a feel for when you'll need to add water. And after a couple of years a little built-in alarm will probably go off in your head whenever the batteries might be thirsty.

A procedure as simple as pouring a little water down a hole should not need much explanation, but a few pointers will get you started on the right foot.

WILLIE'S WARPED WITTICISMS

If dogs had batteries they'd have to be maintenance free. Otherwise they'd slobber away all the electrolyte.

- **Use only distilled water.** Any other type of water will contain minerals that can dilute the electrolyte and collect on the plates, reducing a battery's effectiveness.

- **Bring the battery to a full charge before adding water.** Why? Because the electrolyte expands as the state of charge is increased. If you fill a battery with water and then charge it, acid will dribble out from under the caps. This creates a smelly, corrosive mess and also dilutes the electrolyte, since some of it will have escaped. If the battery is greatly discharged and the plates are exposed, cover the plates with water before charging.

- **Don't overfill the battery.** Adding water to a level just below the bottom of the fill well is sufficient.

- **Never let the water level drop below the tops of the plates.** Exposed plates quickly begin to corrode. At a bare minimum, the plates should always be covered by at least ¼-inch of water.

- **Don't wear loose jewelry** that might cause a short between terminals of opposing polarity. It could be more excitement than you bargained for.

If you can find one, buy a half-gallon battery filler bottle with a spring-loaded spout that automatically stops filling when the level reaches about one inch from the top.

And by the way; if anyone has ever told you that you will dilute the electrolyte by adding water, don't believe them. The only gases that escape a vented battery are hydrogen and oxygen, the constituents of water.

The Battery Box and Keeping Your Batteries Warm

It is neither necessary, nor desirable, to store your batteries outside. The optimum temperature for most batteries is approximately 75 to 80 degrees Fahrenheit. Efficiency falls off as they become colder, and outgassing increases as they become warmer. So, short of a climate-controlled room, the best place to keep your flooded lead-acid batteries is in a box within your house.

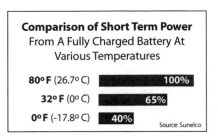

Comparison of Short Term Power
From A Fully Charged Battery At
Various Temperatures

80° F (26.7° C)	100%
32° F (0° C)	65%
0° F (-17.8° C)	40%

Source: Sunelco

The box doesn't need to be fancy: I built both of mine from ½-inch CDX plywood. Many people use large plastic tool boxes. The only requirements are that the box be sealed, vented to the outside, and **not** placed under the inverter (according to the NEC). Door and window weather stripping works fine to seal the lid, and 1-inch PVC

Venting

You do not need a fan to vent your battery box. Hydrogen gas is even lighter than helium, and can easily find its own way out providing, of course, the vent exits the box at the highest point.

pipe at the highest point on the box is sufficient for the outside vent. A nylon screen covering or a plastic kitchen scrubber stuffed loosely inside the outdoor vent opening will keep little furry varmints from exploring the inside of the vent pipe.

If the box is going to be placed on a concrete floor, it's a good idea to support it with treated 2 x 4's. This will allow air circulation under the box and avoid rot.

Wiring the Batteries

Note: The NEC (National Electric Code) does not allow welding cable to be used for inverter cables and battery connections, though many individual inspectors have no problem with it. Before you spend a lot of money, find out what your inspector will accept.

Before you build the battery box, lay out different arrangements for your batteries on paper, because there is more than one way to connect the cables. The object of the puzzle is to keep all the cables—especially those going to the inverter—as short as possible.

The best way I've found is to lay out the batteries in rows, with each row being equal to the number of batteries in a series (2 for a 12-volt system, 4 for a 24-volt system, and 8 for a 48-volt system.) If they will fit the space you have allotted, it makes wiring a simple task.

Let's say you are running twelve 6-volt batteries in a 24-volt system. You will have 3 rows (series) of 4 batteries each. If you lay the batteries out in a 4 x 3 grid, then each row of 4 batteries will be wired in series (positive to negative), to bring the voltage to 24 volts per row. After you do this with each row, you'll find all the remaining negative terminals will be on one side, and all the positives on the other. These terminals will be used to make the parallel connections that combine the amperage from each series (row), without increasing the voltage. Connect all the positives together, and all the negatives, then connect to the inverter with heavy (4/0) cables.

Like water (and some politicians), electrons take the path of least resistance. To make sure all of your batteries are worked equally, you'll need to make the path for all the electrons the same length. How? Take the positive connection from one corner of your battery bank, and the negative connection from the opposite, diagonal corner *(see illustration on the next page)*.

The batteries will require more care and maintenance than all the other components of your wind/PV system combined. Even so, they don't ask for much in comparison to what they give back. And after a few months you'll find that adding a little water and checking the connections now and then is no more of a hassle than taking out the garbage or shoveling snow off the deck (or giving your wife her nightly foot rub).

Just think of your battery bank as your own private herd of short, compact Clydesdales. Batteries may not be as much fun to watch, but you'll get more work out them, and they're a whole lot easier to clean up after.

Step One: Connect each row of batteries in series (positive to negative)

Series Connections increase voltage: add voltage of each battery for total system voltage (6+6+6+6=24 volts)

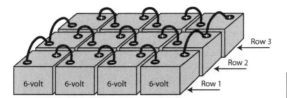

Step Two: Connect rows 1, 2 and 3 in parallel (positive to positive to positive on one side; negative to negative to negative on other side)

Parallel Connections do not increase voltage

Parallel and Series Connections

Connecting batteries in series and parallel may seem confusing, at first, but it's really logical and quite easy. The trick is to make all of the series connections first. That way, the remaining terminals *must* be for parallel connections.

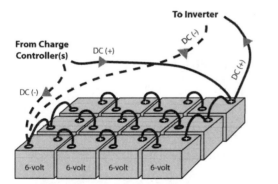

Step Three: Connect incoming current to the bank of batteries, and then run one positive and one negative connection to the inverter. Note the positive (+) and negative (-) cables are on opposite corners to equalize power flow.

Color-Code Your Battery Cables

Knowing positive from negative at a glance is helpful, safe and easy. Before wiring the batteries, lay out all the cables, separating the series cables from the parallel. Wrap a piece of red tape around one end of each of the series cables, and both ends of half the parallel cables. These will be your positive connections. It will help you avoid confusion later, and will probably impress the electrical inspector.

Home-Grown Hydrogen

Will it ever be possible to create hydrogen at home using only sunlight and wind, and then use the hydrogen to run electrical appliances, heat the home, and maybe even as fuel for the family car? The answer is "yes," but it will probably not be practical for a few years.

As you know, current home-based renewable energy systems use solar-electric (photovoltaic) modules and small wind turbines to create DC electricity, which is stored in large banks of batteries. When electricity is needed, the low voltage DC travels from the batteries through a power inverter, where it magically becomes 120-volt AC house current. The inherently low efficiencies of PV modules and wind turbines notwithstanding, the solar/wind-to-electrical-load efficiencies for this type of system are fairly high; in the range of 80 to 90 percent.

In a hydrogen-based system, by contrast, the batteries would be replaced by an electrolyzer, a hydrogen storage system, and a fuel cell stack. By going from electricity to hydrogen, then back to electricity, over 65 percent of the original energy is wasted using current technology, leaving you with a mere 35 percent of the power you started with. So why would you even bother?

A couple of reasons. First, since a fuel cell does not store power as chemical energy, it can deliver power faster than batteries, and will have a much longer useful life. Also, unlike a battery that must keep 50 percent of its power in reserve to avoid damage, a fuel cell will deliver full power as long as there is hydrogen to fuel it, just as a gasoline motor will run full-bore as long as fuel is in the tank.

In addition to electricity, hydrogen can be used for home heating and cooking, and can even be used as car fuel. So how do you make the process less wasteful? You can begin by using a high-temperature fuel cell, such as a solid-oxide fuel cell. By doing this, the excess heat can be used for water or home heating, thus increasing the overall system efficiency.

Even if electrolyzers could someday reach 90 percent efficiency and fuel cells could have a combined heating and power efficiency of 70 or 80 percent, that's still less than batteries. But what if PV modules were twice as efficient and half as costly as they are now? Then suddenly deluxe, home hydrogen-based systems could become even more affordable than today's battery-based system.

Get ready—it's coming. Teams of researchers all around the globe are working furiously to develop low-cost, high-efficiency solar cells. As solar energy grows in popularity this research will only intensify. In the meantime we can look forward to advances in fuel cell, electrolyzer and hydrogen storage technologies. So, when it *does* all come together, it should happen fast.

To learn more about hydrogen technologies and the science behind it, read *HYDROGEN—Hot Stuff, Cool Science (see page 288)*.

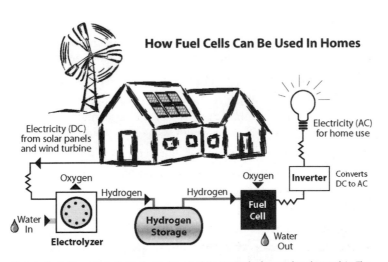

How Fuel Cells Can Be Used In Homes

Electricity (DC) from solar panels and wind turbine

Electricity (AC) for home use

Oxygen

Hydrogen

Oxygen

Hydrogen

Inverter — Converts DC to AC

Fuel Cell

Water In

Electrolyzer

Hydrogen Storage

Water Out

Energy from the sun and wind can be used to create hydrogen by electroylsis. The hydrogen is stored until electricity is needed. A fuel cell then takes the stored hydrogen and adds oxygen from the air to create electricity. Heat created in the fuel cell can be captured for home heating. Water is the only by-product.

Personal Power Companies

Homeowners	John, Kathleen & Ian
# of Occupants	3 adults
Location	Port St. Lucie, Florida
Home Size	2,400 square feet, with a 650 s.f. home office
Home Heating	Alcohol fireplace; electric heat (rarely used)
Water Heating	Water Heating 85 – 90% solar; electric backup
Grid-Tied?	Modified grid parallel (no communication between inverter and grid)
PV Array	540 watts, roof mounted
Charge Controller(s)	OutBack MX60
Inverter(s)	Cotek 1500 sine wave inverter
Batteries	4 AGM sealed lead-acid batteries; 400 amp hours @ 12 volts
Wind Turbine /Tower	None
Solar Hot Water	One 2' x 6' panel, with 80-gallon super-insulated tank. A separate PV module runs the pump
Backup Generator	None
Comments	John's original intent was to keep his home-based accounting business running during lightning-induced power outages, but he has since endured 3 grid-rending hurricanes (Jeanne, Francis and Wilma) with scarcely a blip in his work schedule. In the aftermath of Hurricane Wilma, when they lost grid power for 8 days, John, Kathleen, and their son Ian were hardly inconvenienced; they had enough electricity to keep the home and business running, and plenty of hot water to offer showers to their friends and neighbors.

## chapter 10

Inverters

the last stop on the DC trail

Before it gets to the inverter, the energy that will power your house goes through quite a ride. Energetic electrons knocked out of their orbits by particles of sunlight have charged through the wires at breakneck speed into your batteries, where they were pressed into service to convert lead sulfate into sulfuric acid. The sinusoidal positive/negative waves produced within the wind and micro-hydro generators have been clipped and flipped and transformed into pulsing positive waves of direct current, before joining the current from the solar array in the batteries' chemical energy storehouse. There is enough potential energy sitting in the batteries right now to run your house for several days, but until it goes through one final transfor-

The Mate (left) and the FX2000 inverter by OutBack Power Systems. *Photos courtesy of OutBack Power Systems.*

mation inside the inverter, you can't even use it turn on one measly light bulb.

The inverter is the magic box inside of which DC from the sun and wind (and batteries) is converted into usable AC. Some inverters perform this task better than others.

Sine Waves and Inverters

If Thomas Edison had gotten his way, no one today would be using AC, and inverters would be no more than step-up DC converters. But, since AC has become the norm, sophisticated DC to AC inverters are a must for anyone wishing to live off-the-grid and still use standard electrical appliances.

Alternating current (AC) is delivered in the form of a sine wave. This is a smoothly pulsing wave that gracefully arcs from a peak of positive voltage to an identical negative peak, and back again. Essentially, the current reverses flow with each crest and trough. And it does it very quickly; in the United States, AC and the myriad things that run on it have been standardized to run at 60 hertz, or 60 positive-negative cycles per second.

As we saw in the chapter on wind turbines, a sine wave is the natural form an electrical current takes when it is produced by a coil of wire being rotated between oppositely-charged magnetic poles. But how does an inverter, a solid-state device, take low-voltage direct current (DC)—a flat, old boring stream of electrons—and teach it to do the high-voltage tango?

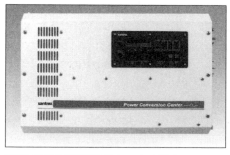

Xantrex Inverter SW4024. *Photo courtesy of Xantrex.*

The heart of an inverter is the transformer. It takes low-voltage DC from the batteries and turns it into the high-voltage AC we use to power our homes. Transformers, however, work on the principle of inductance, a phenomenon that only occurs to any significant degree in the presence of a pulsing (alternating) current. It takes some clever trickery to convince the transformer the direct current driving it is AC.

The magic behind the deception is a configuration called an H-Bridge. Each of the two legs of the H have a transistor switch near each end—four switches in all—and the legs are joined by the transformer in the middle. The two bottom switches control the flow of negative current from the batteries, while the upper switches control positive flow back to the batteries. By electronically timing the opening and

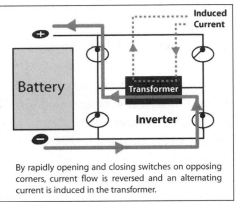

By rapidly opening and closing switches on opposing corners, current flow is reversed and an alternating current is induced in the transformer.

closing of the switches, the current can be made to flow first one way, then the other, through the transformer. Viola—alternating current!

This basic configuration can be used to create modified sine waves, such as those produced by Xantrex DR series inverters. To produce a much better approximation of a sine wave (less than 5 percent harmonic distortion) a series of H-Bridges and transformers of varying voltages are used, creating in essence a series of inverters whose outputs are mixed in ratios determined by the battery voltage.

■ **Sine Wave Inverters**

State-of-the-art sine wave inverters will produce a current as clean as the smoke-belching utility company on the far side of the mountain. If there is anything that won't run efficiently on the current they produce, we have yet to find it. As you might imagine, such technology doesn't come cheap.

We bought a Trace SW4024 sine wave inverter in 1999 when we first moved to the mountains. Initially, we installed it in the cabin, then moved it to the house once we had the rest of system installed. Except for one incident that wasn't really the inverter's fault (I'll explain what happened later) it has performed flawlessly. In the many years of service, the only problems we've had have resulted from my own programming errors, or oversights, rather than the inverter's design.

■ Modified Sine Wave Inverters

For a lot less money, you can equip your house with a modified sine wave inverter, such as one of the Trace DR-series or any of a number of inexpensive, lightweight inverters, but I would advise against it. These inverters produce a stepped waveform that is really just a choppy approximation of a sine wave. For lights and toasters and vacuum cleaners, it may be good enough. But for certain other appliances, problems can arise.

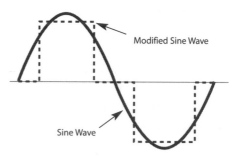

After moving the sine wave inverter to the house, we bought a Trace DR 2524 modified sine wave inverter for the cabin. We were told by some people that it would eat an HP LaserJet computer printer in a hot minute. Others agreed, but said it would take a month or two to do it. I'm happy to say that we're still using that printer, after running it for more than a year with power from our modified sine wave inverter, though there were times when the printer would stall for several moments, before printing. The one appliance it would not run

---

Comparison of Slow Cooker Cooking

LaVonne had noticed a time difference when using a slow cooker (Crock-Pot) to cook a pot of stew with a modified sine wave inverter versus a sine wave inverter. So we tested the usage with a Watts-Up? Meter and found it used the following wattage continuously:

	LOW	HIGH
Sine Wave	178 watts	241 watts
Modified Sine Wave	171 watts	234 watts

After many hours, the difference adds up!

at all was LaVonne's Pfaff serger, which is a fancy sewing machine that cuts fabric as it stitches. Probably it has more to do with the rheostat that runs the machine, than the machine itself.

At the back of manual for the DR series inverters, Trace has compiled a list of devices that may experience problems running from a modified sine wave inverter. These include (but are not limited to) microwave ovens, clocks, ceiling fans, dimmer switches, and rechargeable devices. I quickly discovered that I could not let my DeWalt 12-volt drill batteries sit in the charger for very long after they were charged, or they would begin to overheat.

The bottom line? If you are installing solar electricity in a weekend cabin, a modified sine wave inverter will probably work fine. But for a house with all the modern conveniences you have come to rely on, don't scrimp: a good sine wave inverter is the only way to go.

Inverter Functions

What else can an inverter do, besides change low-voltage DC into usable AC house current? Plenty. A top-of-the-line inverter will have more features than you will ever use. Here are a few to look for:

■ **High and Low Voltage Shut-Off**

A high and low voltage shut-off is the inverter's way of protecting your appliances, the batteries and—most importantly—itself. There should be one programmable setting for low voltage shut-off, and another for high voltage shut-off. The factory defaults are probably fine, unless (as previously mentioned) the high voltage shut-off is set lower than the voltage allowed by the charge controller during equalization. If this happens, the inverter will turn itself off.

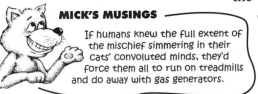

MICK'S MUSINGS

If humans knew the full extent of the mischief simmering in their cats' convoluted minds, they'd force them all to run on treadmills and do away with gas generators.

■ Battery Charging

You will also want an inverter that doubles as a battery charger. It should have settings for bulk and float voltages, and they should be set the same as on the charge controller.

■ Generator and Grid Tie-ins (for systems with batteries)

If you are completely off the grid, then the second feature (grid tie-in) will be of no interest to you. But if you are connected to the grid and still have batteries, a good sine wave inverter will stay in sync with the outside power source, cutting in when there is a power outage, or, depending on where you live, selling power back to the utility when there is an abundance.

Off-the-grid homes need a tie-in for a generator. Some inverters will start the generator for you—providing the generator is wired for remote start—when the batteries get too low (though you really should size your system so the batteries don't often get so low). Mostly, you'll use the generator after a few cloudy days when you want to run a heavy load, like a washing machine, or a dishwasher. One thing to watch out for: make sure the "maximum charging amps" setting on the inverter is within a range that the generator can handle; otherwise it will trip the breaker on the generator. Even if it doesn't trip the breaker, the inverter might draw the generator voltage down below an acceptable level, and disconnect itself. Something to keep in mind, so you'll know what happened, if it happens.

■ Search Function

The search function is a feature we use at the cabin, but not at the house. It is designed to save power, but it can cause problems. Essentially, the inverter in search mode is at rest, but it sends out a pulse of current every second or so (this pulse-rate is adjustable) to see if anything gobbles it up. If it does—such as when you turn on a light—the inverter will come to life and power the load. This is all well and good, until it finds a load that takes more than the preset search wattage to start, but less than the search wattage to run.

Usually, this is just annoying. But if you are not careful, it can get expensive and even dangerous. For example, I was charging a DeWalt drill battery with the inverter set in search mode. (This is a different story than the previous one. Most of my inverter problems seem to revolve around drill batteries.) As long as the battery was charging there was no problem. After it finished charging, however, and the inverter went back into search mode, the battery charger interpreted each pulse of current as a signal to start up again and sent a surge of power to the drill battery. Then, sensing that the battery was already charged, it would shut down, only to go through the whole cycle again. The result was a $50 battery bursting at the seams, and too hot to handle.

Once we moved to the house, we simply left the inverter in the "on" mode. At absolute rest, our house consumes 45 watts of power. The loads include: the inverter itself; three smoke detectors (required by the county to be hard-wired into the power supply); the displays on two charge controllers and two TriMetric meters; and the clocks on the gas range and the microwave oven. All other appliances that might draw a ghost current (television, satellite receiver box) are plugged into surge protectors with disconnect switches.

■ Stackable Inverters

No matter how big you think your inverter is, make sure it can be "stacked" with another, identical inverter. If your house has a number of heavy loads that might run simultaneously, or if you one day discover that the inverter you bought is too small for your growing needs, it may be wise to get two inverters and stack them, so that they operate in phase, as a single inverter. The inverters may be wired in parallel, providing twice the amperage at the same voltage (120 VAC), or they

---

Other Functions of Inverters

■ High and low voltage shut-off
■ Battery charging
■ Generator and grid tie-ins
■ Search functions
■ Stackable (for greater voltage or amperage)
■ Computer interface

---

can be wired in series, to double the voltage (240 VAC). A transformer can be used to step-down, or step-up, the voltage for certain loads, such as a 240-volt well pump.

With renewable energy growing in popularity, people are building bigger houses with more appliances that can often tax the resources of a single inverter. For that reason, Xantrex Engineering and others now manufacture power panels. These are pre-assembled units with one or two inverters and charge controller(s), along with a DC disconnect, transformer, and whatever else you may require. As you might imagine, power panels are expensive, so shop around for the best price.

OutBack Power Systems now has four inverters on the market, ranging in output from 2,500 to 3,600 watts. OutBack uses the concept of an inverter module, which means that the programming is not done on the inverter itself, but from a remote display unit called the Mate. Multiple inverters can be stacked without you having to shell out money for the display and programming electronics for each one, which is nice. And with all the programming being done on one remote unit, there's no chance of conflicts between different inverters. This is good, because no one wants their inverters locked in eternal battle with each other.

Power Conditioning Center by OutBack.
Photos courtesy of OutBack Power Systems.

OutBack's PS1 Power Center,
as seen with x-ray vision.

OutBack also offers a PS1 Power Center for those of you hooked to the grid but needing a little backup now and then. The PS1 combines a 3,000-watt inverter and an MX60 charge controller along with all the requisite safety equipment in one tidy outdoor enclosure. Beneath it is another enclosure designed to hold four 100-amp-hour 12-volt sealed batteries. The entire unit takes up about as much room as a utility worker leaning against a wall. The PS1 won't run your critical systems for days on end, but can be enough to get you from one small to medium power outage to the next.

■ Computer Interface

Most high-end inverters can be monitored and programmed via computer, as long as you have the interface and the software to run it. I haven't tried it myself, mostly because I really enjoy accidentally pushing the wrong button and shutting down the whole house once in awhile. Still, I'm sure it would be handy to have sometimes; it's got to be better than resting on your knees on a concrete floor, pushing buttons with the hand that isn't holding the user's manual.

I know I've said it before, but I'm going to say it again anyway: don't scrimp on the inverter! For a full-fledged house, a programmable sine wave inverter is a must. Get a big one, if not two. That way, when the well pump is pumping water for the washing machine, you can still run your table saw without causing your own personal power outage. It will save you a few moments of button pushing, *after* several minutes fumbling around in the dark trying to find a flashlight. And it may spare you a derisive comment or two.

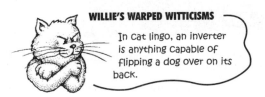

WILLIE'S WARPED WITTICISMS

In cat lingo, an inverter is anything capable of flipping a dog over on its back.

Grid-Tie Inverters
(for systems without batteries)

The inverters used for direct grid intertie systems are a whole different breed of animal from the multi-purpose inverters used by the rest of us. A direct-tie inverter wouldn't know how to charge a bank of batteries even if you wanted it to, and there's no place to hook up a gas generator. If the utility grid goes down, the inverter goes down with it, depriving you of any power you might be producing from your solar array and leaving you without any means of backup power.

Direct-tie inverters are designed to do just one thing and they do it quite well. During the day when your solar array produces more wattage than your home is using, the inverter will power your home from the array and sell any excess back to the utility. At night, when the array is idle, the inverter will prevent any grid power from trickling into the array.

While dispensing with all the circuitry devoted to battery charging and generator tie-ins, direct intertie inverters also do away with the need for separate charge controllers. Using maximum power point tracking (MPPT) technology, these inverters take direct current (DC) from 48 up to 550 volts, depending on the brand, and efficiently convert it to usable AC amperage. And, since they are designed to tie directly into the power grid, these inverters are generally made to be mounted outside the house.

Because battery-less systems are becoming so popular in certain places, there are a number of good direct-tie inverters on the

A Sunny Boy 2500 is mounted on a north-facing outside wall, and a standard SquareD DC disconnect is mounted to the right of it. *Photo courtesy of Summit Electrical Service, Santa Rosa, California.*

market. Sunny Boy inverters from SMA America got into the game at the right time and have held center stage ever since. SMA America offers five inverter models ranging from 700 to 6,000 watts, with the Sunny Boy 2500 being the most popular.

For direct-tie wind systems, SMA has come out with the Windy Boy line of inverters. At present there are two models, inputting either 1,800 or 2,500 watts. These units are only sold as a package deal that include the wind turbine (not manufactured by SMA), since each model of turbine requires slightly different programming.

Other direct-tie inverters that are daily proving themselves out in the field include the Fronius IG-series; the Sharp SunVista inverter; PV Powered inverters; and Xantrex, which has made it back on the direct-tie scene with their GT 3.0 inverter. You will find website listings for all these inverters in the appendix. Check 'em out. They each come with their own distinct features (too numerous to list here), so if you're in the market for a direct-tie inverter you shouldn't have much trouble finding one to fit your needs.

For more detailed information on grid-tied systems, see the book that I co-authored with Doug Pratt: *Got Sun? Go Solar*. A look at the advertisers in *Solar Today* or *Home Power* magazines will keep you abreast of new manufacturers and models.

Personal Power Companies

Homeowners	John
# of Occupants	1 adult
Location	Loveland, Colorado
Home Size	4,000 square feet on 2½ levels
Home Heating	In-floor hydronic; plus wood stove
Water Heating	Propane-fired boiler (and future solar hot water)
Grid-Tied?	No
PV Array	1,920 watts in 2 ground-mounted arrays which are 100 feet south of house
Charge Controller(s)	OutBack MX60
Inverter(s)	Xantrex SW4048 sine wave
Batteries	16 Trojan L-16H 6-volt flooded lead-acid; 780 amp hours @ 48 volts
Wind Turbine /Tower	None
Solar Hot Water	Piping roughed in, but panels not installed
Backup Generator	Kohler 8.5 kW propane-fired; wired for automatic start
Comments	John's system is expertly installed and runs trouble-free, 4 miles from the nearest power pole. After the first year he added 480 watts of solar and now spends far less time listening to his generator during the cold Colorado winter. John saves watts by using a Servel propane refrigerator.

## chapter 11

Putting It All Together — Safely

protecting you and your system from each other,
and from nature

Imagine that lightning never wandered beneath the clouds. And while you're at it, imagine that you could be absolutely assured that every part of your system would work perfectly at all times—never requiring service or replacement parts or components—and that no one would ever try to draw more current through a wire than the wire could safely carry. If these three conditions were always true, then there would be no need for fuses, breakers, disconnects, ground wires or lightning surge protectors.

But this is planet earth, we're all humans here, and even when nature is on her best behavior components still wear out and we still end up doing things with electricity that we shouldn't, though we claim to know better. Since this is the way of the world and there's nothing we can do about it, we—sentient beings that we are—can, and most assuredly should, take precautions against the inevitable.

In this chapter I'll discuss grounding, over-current and lightning protection, and how important each procedure is to your home. I will then go through a typical system, beginning with the solar array, wind turbine and micro-hydro generator, and ending with the inverter. At each step along the way, I'll describe the minimum precautions you should take to protect your system and yourself from the forces of nature, and any inherent dangers that might be present when replacing, repairing, or reworking any part of the system.

These recommendations are sound practices that have worked for me and other people I know with our particular systems—and the two or three by-the-book electrical inspectors that roam these parts. Your system (and maybe your inspector) may differ in subtle, but significant ways, which means that your system might require additional safeguards that are not mentioned here. As in all things electrical, the inspector has the final say. Nor is the National Electric Code a static thing. It grows, every year. What passes code today may not pass code tomorrow. What passes in one state or county may not pass in another. That's why electricians get the big bucks and the homeowner wires his house at his own peril.

It is not especially difficult to wire a renewable energy system, but it does take a clear, logical mind, an appreciation for detail, and a lot of time. If you think you can put together your own system, you probably can. If you have doubts, it might be better to leave it to a professional.

This chapter is intended to be a discussion of safety components, and where they should be located in relation to the working parts of your system. Logistics, in other words. If you are hoping for specific information on wire sizes and types; conduit sizes and types;

WILLIE'S WARPED WITTICISMS

An electrical shock may put a tingle in your teeth, but a good cat scratch will show you what you're made of.

fuses, breakers, boxes or fittings, don't hold your breath, 'cuz you ain't gonna find it here. There's simply not enough room to cover that much ground in so few pages. Besides, it's boring stuff.

Should You Install Your Own System?

In case you haven't guessed, I'm a hands-on, do-it-yourself guy. I can't bear to pay someone to do something I can do myself. It's an attitude that often gets me into trouble but, hey—that's me. So it may seem disingenuous of me to suggest that it might be in your better interests to hire someone to install your system. After all, I'm sure you know your own limitations better than I do.

But if you are planning a system that hooks into the power grid, I seriously suggest you consider having it done by a professional installer. Why? Two reasons. Number one, you'll be dealing with lethal doses of electricity; if you make one mistake you may not be around to make a second one. And number two, many rebate and tax credit programs carry a self-install penalty. In other words, if you do the work yourself you'll feel it in the wallet. It's a great incentive to kick back and watch someone else do the work.

Grounding

The purpose of grounding a system is to provide an alternate path for current to flow, should a current-carrying conductor (the positive lead from a solar array, for instance) ever come into contact with a metal surface (such as a fuse box) that you might touch. With a good, unimpeded path to the earth, where the charge can quickly dissipate, the current won't have to try and seek ground through you, because, unless you're wearing chain mail, you are not as good a conductor as copper wire.

On the AC side of your system, the (white) neutral wires, and the (green) ground wires will all ultimately lead to a copper ground rod outside the house. In turn, every light fixture, outlet box, fuse box, junction box, and PV/wind/hydro component encased in metal should have a path to ground. Any good electrician will know all about proper grounding of the AC side of things.

Grounding of the DC side is similar. All the negative leads from the wind and solar sources will ultimately be connected at the batteries, and all should have a path to the same ground rod as the AC side.

■ Bonding

The point where the AC neutral and ground wires join with the DC negative lead is called the point of bonding. This should be done at exactly one point within the system. It may be done at the ground rod or the inverter, but usually it is done at the AC service panel. If you install a bypass switch (to run your house from grid or generator power in the event your inverter fails) it can also be done there. All that matters is that it is done *somewhere*. Local codes may vary on the point of bonding. To be safe, explain to the electrical inspector that you wish to bond the DC negative to the AC neutral and ground leads, and then do whatever he or she suggests.

Lightning Protection

When preparing for the dangers of lightning, it's important to understand that lightning *wants* to go to the ground, so anything above ground is subject to its effects. Solar arrays are particularly vulnerable. During one especially worrisome storm at the cabin, our charge

One of six lightning rods on our roof.

controller went into overload protection mode three times, due to lightning hitting our array. Fortunately, the only damage was a few frayed nerves.

Lightning occurs when there is a massive disparity in charge between the clouds overhead, and the ground. The idea of a lightning rod is to provide a path for positive charges from the ground to cancel out the negative charges in the air. In that way, a lightning rod helps to keep lightning from occurring in the first place, since it serves as a bridge between the sky and the ground where the needs of the former can be offered up by the latter. But if the clouds demand more than the ground is willing or able to give, then a lightning rod provides a path to ground that does not run through your house.

In many ways, solar arrays and wind towers act as lightning rods. This means that they help equalize the electrical potential that exists between the ground and the sky. It also means that they will both attract lightning if the electrical potential becomes too great. There is nothing you can do to change that fact. But there is a lot you can do to mitigate the effects of lightning.

■ Well Casings

Heavy metal well casings are notorious for attracting lightning. If lightning is a problem in your area, you may want to consider asking your pump installer to wire a lightning arrestor into the system to help prevent a damaging surge of electricity from zapping your inverter.

■ Log Homes versus Framed Homes

For those of you living in, or planning to build, a log home, lightning protection is a must in lightning-prone areas. This is one of the few things we found out the easy way. Being on top of a hill, we thought it would be a good idea to have lightning rods installed on our log house. Since lightning protection is something of a secretive art, we hired a professional installer to ensure that it was done right.

When he saw that we had a log house, he told us how wise we were to install lightning rods, since log houses always sustain more damage from lightning hits than conventionally-framed houses. Often, he said, a log house will burn to the ground, from the excess heat it absorbs from a strike.

Since logs are laid horizontally, rather than vertically, a log wall presents a difficult path to ground. This means that lightning has to try a little harder. Bad news for the house.

The Solar Array

■ Fuses and Breakers

Each solar module is designed to carry only so much amperage. The bigger the module, the more amperage it can handle. When you wire modules in series you are increasing the voltage, not the amperage, so you can safely wire two or four modules in series without concern for overloading the module wiring.

An OutBack combiner box with 12 circuits, and a lightning arrestor in the lower left corner. *Photo by Doug Pratt.*

But as soon as you wire one series of modules to another series of modules in *parallel*, then you *are* increasing the amperage, and it is imperative that you isolate each individual series with a fuse or a breaker to protect it from a reverse electrical surge, as could happen with a short circuit somewhere in the line.

Typically, a combiner box is used for this purpose (OutBack Power Systems makes a good one). The leads from a series of modules enter the box and run through a fuse or a breaker, before its amperage is joined with the amperage from other series of modules.

In addition, there should be one common disconnect for each array *after* all the current is flowing into a common feed. This is so the array can be shut down quickly and easily, either because of an emergency, or simply to service "downstream" components. A breaker box with a single properly-sized, DC-rated breaker works well for this purpose.

■ Grounding and Lightning Protection For Your Solar Array

Note: *At the time our PV/wind system was inspected, the National Electric Code had no regulations regarding lightning protection. I have heard rumors this is due to change. If it does, any recommendations made here could be rendered moot.*

The array must be grounded. Period. Every module to every panel frame, every panel frame to the entire array, the array to a copper ground rod via a heavy copper ground wire. Obviously, the shorter the route to ground, the better. A buried ground rod beside the array and connected to the common house ground is ideal.

The grounding system and the system's fuses and/or breakers will protect your array from electrical accidents or oversights, and most ambient electrical surges from nearby lightning. If you have a good charge controller, it will disconnect from the array, even before the fuses and breakers can react. For more energetic strikes, a lightning arrestor is a very good idea. This is a small device that "absorbs" excess voltage, then slowly dissipates it. It should be mounted close to the array; either on the combiner box or on the main disconnect. Most have three leads: positive (red), negative (black), and ground (green). The ground wire may be connected to the disconnect box (if it's metal), and the box should have a path to the heavy copper ground wire running from the array to a common ground.

The Wind Turbine

■ The Wind Brake

Whether your wind turbine sends AC or DC to the house, the electrical inspector will want some sort of disconnect outside the house. This will be a switch that stops the propeller from turning (a wind brake), and may or may not be supplied with the generator package. If it isn't, you can order it separately.

■ Grounding and Lightning Protection for Wind

It may seem that a wind tower, set in solid concrete and guyed to the ground with heavy steel cables, would be well grounded. Unfortunately, it may not be. Since your wind tower will be the tallest thing around, you should take extra care to protect it from lightning.

To help avoid a lightning strike—and to minimize the effects, should one occur—drive a copper ground rod next to the tower and connect it to the tower with heavy copper ground wire. To be *really* safe, drive one ground rod next to the tower pad, and another rod at each of the guy wire pads. Connect all the ground rods together with heavy copper wire (#6 or bigger), connect them to the tower and each of the guy wires, and then run a ground wire to the common house ground. Bury all the wires at least 6 inches below the surface. Use heavy copper connectors, and make sure there is good contact (no rust, or paint). Expensive? In the grand scheme of things, not very. Worth it? If it's needed even once, yes.

A lightning arrestor should also be installed at the outside disconnect box, to help disperse any excess voltage that manages to get inside the lines. Since wiring differs from one type of wind generator to the next, you should ask the manufacturer what type of lightning arrestor (about $40) to buy. Most likely, they will be happy to sell you one. Buy it.

Micro-Hydro Safety

Your micro-hydro system should not require any special grounding or safety features beyond those already mentioned for solar and wind systems, though if the turbine is mounted on a metal frame it should be properly grounded.

Instead of an electrical or mechanical brake (as for a wind turbine), or a circuit breaker (like with a solar array), you will use the ball valves located near the nozzles to stop the turbine from spinning whenever you need to shut off the power. How well this flies with the electrical inspector (whose job it is to inspect electrical stuff, rather than plumbing stuff) is anybody's guess, but once you explain (a) it's your only option; (b) it really does shut down the power source; and (c) people use these systems everyday (even if your particular inspector has never seen one), I can't imagine you'll have much trouble.

Charge Controllers

With outside disconnects for both the wind and solar charging sources, there need be no other breaks in the lines, until the current moves past the charge controller(s). There should be some type of over-current protection between the charge controller(s) and the battery bank. A properly-sized inline fuse will work, though a DC-rated breaker will make it easier to isolate the battery bank from the DC sources and is more likely to bring a smile to the electrical inspector's face. (And we want the inspector to smile, don't we?)

Some wind charge controllers come with their own built-in wind brakes. It is less common for solar charge controllers to have built-in disconnects. Plan accordingly. The breaker, or fuse, should be rated for slightly more amperage than the surge amps the charge controller is designed to handle.

DC Disconnect

The DC disconnect is usually a very large DC-rated breaker that lies between the battery bank and the inverter. It is designed to protect the batteries—and the inverter—should a short circuit occur somewhere within the battery bank. It also makes it very convenient for shutting down power to the inverter, and therefore the entire house, whenever you need to. The DC disconnect is sized for the inverter, and is designed to trip only when it senses far more amperage in the lines than the inverter could ever hope to use.

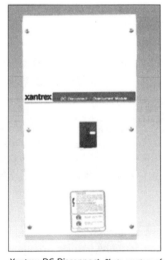

Xantrex DC Disconnect. *Photo courtesy of Xantrex.*

Xantrex makes a 175-amp and a 250-amp DC disconnect. Both can handle more than their rated amperage for a short time, so they don't trip when the inverter surges to higher amps. Our Trace

SW4024 inverter has never tripped our 250 amp DC disconnect, though it has certainly drawn more than 250 amps from the batteries for very short intervals. Because it's directly connected to the batteries, the DC disconnect is the favored place to connect all incoming and outgoing DC circuits. A good DC disconnect will provide mounting space for additional smaller breakers. The Xantrex disconnect provides four spaces, the OutBack disconnect provides ten.

WILLIE'S WARPED WITTICISMS

Dogs would be far more manageable—but considerably less exciting—if they all came equipped with disconnects.

Wiring it Safely

With properly placed breakers and disconnects, you will be able to wire most of your system without the threat of electric shock, or a component-destroying short circuit. The solar array and the batteries are two exceptions. To avoid the possibility of damaging the solar array by accidentally crossing the wrong wires, it's a good idea to cover the array with a blanket or tarp before trying to perform the delicate task of wiring the modules together in series and parallel.

The batteries are another matter. By nature they are full of electrical potential and need to be wired with the utmost care. Know what you are going to do before you do it. Never work on the batteries with-out first isolating them from every other component within the system. This means flipping the disconnects (breakers) that

---

DC Wire Coding Colors

POSITIVE = Red
NEGATIVE = Black

NOTE: *The NEC now states that* **black=positive** *and* **white=negative** *on DC systems, though in practice, this new coding is rarely used on DC-AC systems.*

To avoid confusion, try using **white** *for negative (DC) and neutral (AC) connections, and* **red** *for positive (DC) and hot (AC) connections.*

AC Wire Coding Colors

Red or Black = HOT
White = NEUTRAL
Green = GROUND

lead *from* the charge controller, and *to* the inverter.

There are basically two ways to coax a spark out of a battery: shorting together the positive and negative terminals on a single battery (wrenches and screwdrivers are the usual culprits), or shorting the positive and negative leads from a series or parallel string of batteries (usually this requires carelessness or confusion).

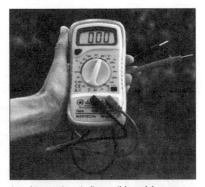

A multimeter is an indispensible tool that can test for AC and DC voltage, DC amperage, continuity, and resistance.

Know which cable is which! Color code all the cables with tape (positive = red; negative = black) before connecting them—either to the batteries, or to the terminals of the charge controller, breaker boxes, or disconnects. Draw yourself a wiring diagram before you begin and consult it every step of the way. When finished, tape it inside the battery box for future reference. It may be that you know exactly what you are doing at the time you're doing it, but come back a few months later, and any wiring without color coding is going to look incomprehensible, especially on the AC side (where black = hot, white = neutral, and green = ground). At that point, you have to pull out the multimeter to decipher what you did.

A friendly word of advice for procrastinators: hide the multimeter under rolls of red, black, green and white tape. The tape will remind you why you are looking for the multimeter in the first place.

MICK'S MUSINGS

If cats were color-coded, most would be red. Willie would glow like Chernobyl's reactor.

## chapter 12

Heating of House and Water

staying warm without busting the energy bank

This chapter will be most helpful to those of you who are currently planning a new home powered by renewable energy sources. If you intend to install a PV/wind/hydro system in an existing house, you won't have much choice but to adapt it to the existing heating system. Fortunately, most heating systems will work with renewable energy sources, though obviously some are better than others. (On the other hand, if your current home uses electric heat, you should either sell this book to someone else, or buy or build another house. I strongly endorse the latter option.)

By the time LaVonne and I began planning the new log house, I had spent seven years of my life living in a cabin with no heat other than a wood stove. Before and after the cabin interlude—except for the years spent in my parents' house, which was heated with hot water circulating through baseboard registers—I lived in dwellings with forced air (a.k.a. forced dust) heating systems.

Between the two, I far preferred the wood stove; it was quieter, easier to control, and cheaper to operate. But, after an absence of a day or two in the middle of January, it took a steely constitution to come home in the dead of night.

Though LaVonne, poor girl, missed out on the earlier cabin years—chopping holes through the ice in the creek for bath water is one my fondest memories—she was all too familiar with forced air heating. We were united in our loathing for it. Owing to that fact, we were able to quickly focus our heating options when it came time to build the house we'd been dreaming about since our courting days.

Obviously, electric heat was out; the PV/wind system needed to run it would rival the cost of the house. Propane wall heaters are a good, cheap, solution for cabins but quickly lose their practicality in large, multi-room houses. That left us with the hot water option. Baseboard or in-floor? It was a no-brainer: definitely in-floor.

But I'm getting ahead of myself. Before I let *all* my biases dangle in the wind, I'd better do a little explaining.

Propane Wall Heaters

Propane wall heaters are small, inexpensive units that can be mounted on a wall. They can be purchased either with a blower fan, or without one. As backup heat for a cabin or small house with a wood stove, they're terrific. We have several friends who use them. The heaters save them the trouble of having to stoke the fire in the middle of the night, and keep their houses adequately warm if they're away for extended periods. We installed one in the (usually vacant) cabin last year to keep the pipes from freezing on cold, cloudy days in the dead of winter. It's saved me from many cold walks to stoke the wood stove.

But a propane wall heater is really pretty much a brainless animal, with little hope of ever getting any smarter. It senses how cold it is in one location, then heats that location until it decides it's warm enough. It doesn't give a hoot how much heat makes it to the bathroom, or the far bedroom, which is unfortunate, because it's doubtful the building inspector would allow one to be mounted there to keep the chill out of your bones during frosty, winter nights.

So, for a very small house on a no-frills budget, a propane wall heater may suffice, but for a larger, comfortable house, you will need something more.

Forced Air Heat

Most houses in America—both old and new—are heated with warm, forced air. The air is warmed in a furnace (usually gas-fired), then pushed through a labyrinthine system of ductwork with the aid of an electric blower (a squirrel-cage fan) into each room of the house through vents cut into the floor or ceiling.

It's fairly inexpensive to install, and we know people who use this system in their PV/wind homes with reasonable success. But we don't know anyone, off the grid, or otherwise, who *likes* it.

Energy-wise, forced air heat is less efficient than hot water heat, requiring more propane or natural gas *and* more electricity to heat a house. This is mostly due to the fact that air is thinner than water. It goes places we don't want it to go, and requires more energy per unit mass to heat.

Many of the problems with forced air heat are due to poor system design. Anyone who has ever lived in a house heated with forced air knows how easy it is to throw the whole system out of kilter, just by shutting a door or two. Some rooms get too hot, while others get colder. Often the only way to achieve any degree of equilibrium is to leave open all the doors to all the rooms—not always a satisfying solution.

This problem can be alleviated somewhat by providing properly placed cold air return ducts, so the air can re-circulate

Geothermal Heating and Cooling

Ground-Coupled Heat Pumps: Large-Scale Refrigeration

Geothermal systems (also called ground-coupled heat pumps) circulate fluid through underground pipes where the Earth is a constant temperature...about 50 degrees Farenheit. This fluid is then converted into hot or cool air for the home, as it is needed throughout the seasons. These systems are becoming more popular with grid-connected homeowners as an efficient and cost-effective method of heating and cooling their homes.

Like refrigerators, heat pumps are able to capitalize on the temperature differentials between a refrigerant in its gas, liquid, and compressed liquid states. As cold refrigerant comes in contact with a loop of antifreeze solution moving through a series of pipes buried in the ground outside your house, it absorbs heat. Once the heated gas reaches your house, a compressor forces it back into a liquid state, concentrating the heat within the gas which is then used to warm the house. Having dumped its heat load, the compressed liquid passes through an expansion valve where it is once again transformed to a cold gas, ready to absorb more heat from the ground. You can think of the ground as the heat source, and your house as the heat sink. In summer the process is reversed: the ground, which had been greatly depleted of its heat during the winter, now becomes the reservoir for the excess heat absorbed from your house.

It's a great idea that requires no on-site fuel such as propane or natural gas. But like refrigerators or air conditioners, ground-couple heat pumps do take a fair amount of electricity so they really won't work well with off-grid systems. However, if you have a small acreage (or even a big yard) with good soil in a location that offers incentives for grid-tied systems, a geothermal system just might be the answer. To get started learning about this interesting concept, visit the NREL website at: *www.nrel. gov/learning/re_geo_heat_pumps.html.*

throughout the house without building up pressure gradients in certain areas, pushing warm air out of the house in some places, and drawing cold air in, in others. But, as long as the entire house is run off a single, centrally-located thermostat, the problem of uneven heating will persist.

Zoning is the obvious solution. By dividing the house into three or four distinct zones, each on its own thermostat, forced air heat becomes almost comfortable. Zoning comes with a price, however. Besides upping the original cost of the system by at least a couple thousand dollars, each zone requires a small motor to operate a zone damper within the duct, in addition to the main 5-amp (600-watt) blower motor located inside the furnace. It can add up to lot of wattage at the time of year you can least afford to use it. And you'll still have to deal with dry air blowing up through the floor at incalculable intervals, a noisy blower, and severe limitations on how you can arrange your furniture.

Of course, any system that requires fans or pumps will tax your electrical energy budget; there's no way around it. But, if there were a system available that used less electricity and gas; a system that took up zero floor space while providing even, comfortable heat, wouldn't it make sense to use it?

Heating With Hot Water

Hot water heat has been around in one form or another much longer than natural gas has been available to fire the boilers that produce it. Even the Romans used it to heat their floors. And today we still see cast iron radiators in many older houses, school buildings, and court houses. A lot of those old coal-, wood- and oil-fired systems are still in use (now mostly converted to natural gas or propane), but their day is done; evolution has taken its course. Today's hot water heating systems come in two basic incarnations: baseboard registers and in-floor radiant heat.

Of the two, in-floor radiant heat provides the most even heat throughout the house. We chose it because it was the most logical choice for a log home, for the simple fact that log walls—being round by nature—do not have flat baseboards to which you can attach the registers. (Okay; I'll concede that any enterprising home builder could find a way to make a flat, vertical surface along a log wall beset with bumps and knots, but I defy anyone to make it look like it belongs there.)

The fact is, if you install radiant floor heat in your home—log, frame, straw bale, cinderblock, adobe, beer can or waddle and daub—you won't be disappointed. I believe it's the perfect way to heat any home, for reasons that will quickly become apparent.

■ Radiant Floor Heat

The idea of radiant floor heat is to make the entire floor one huge wall-to-wall radiator. To accomplish this, a continuous length of extremely tough PEX (cross-linked polyethylene) tubing is snaked back and forth across the subfloor, stapled down, then embedded in a special type of lightweight gypsum concrete that is poured over the floor in a soupy slurry that hardens in two or three hours.

You can have as many or as few zones as you need, without need of a nightmarishly complex system of ducts. Each zone runs on its own thermostat, which controls the pump for that particular zone. We have five zones in our house: one for the garage, and another for the workshop and electrical room; on the main floor we have our great room on a zone, and the office and bathroom on another. The pantry is unheated. A single zone in the loft heats the upstairs bathroom. The rest of the open loft stays plenty warm from the heat rising from below.

Since we heat the great room and kitchen primarily with a wood stove (whenever we're around to stoke it) we set the thermostat lower there than in the office and bathrooms. It's a very comfortable house, even during the coldest winter nights.

Special Considerations for Radiant Floor Heat

An oft quoted limitation of radiant floor heat is the amount of time it takes to warm a room, once the thermostat is turned up—there is no instant heat, as there is with forced air. This is because no additional heat can be felt until the hot water running through the PEX tubing raises the temperature of the medium in which the tubing is embedded.

If you lived in a drafty, poorly-insulated, stick-built house (of which there is no dearth in this country) this could be a real problem. But you don't. Or at least you won't, if you are building or planning to build a house. With modern windows and highly insulative building materials, a well-built house will retain heat so efficiently that it is doubtful you will even know when the heating system is working and when it is idle.

Log homes are especially suited to radiant floor heat because of the tremendous thermal mass of logs. When the house is warm, the log walls soak up heat, then release it as the inside air temperature cools. In effect, the walls of a log house act as heat radiators, just like the floor.

You will be limited, however, in your choices for the finished floor. Carpet is feasible, though a thick, plush carpet above a standard, airy carpet pad won't work very well. You'll need a dense rubber carpet pad (not foam) to allow for the passage of heat. Considering the growing popularity of radiant floor heating, any reputable carpet distributor should know what will work and what won't.

Helpful Hints When Installing Radiant Heat In New Construction

- Increase your ceiling height to allow for the thickness of the Gypcrete®.
- Double plate the bottoms of your framed walls.
- Make sure your floor joists can support the extra weight.

If you elect, instead, to go with a wood floor, you'll need to install an engineered floor, one that "floats" on the surface of the gypsum concrete, with no points of attachment. A floating floor can expand and contract, without warping or buckling, as it heats up and cools down. There are a few brands of engineered floors (not the thin, laminated variety) on the market; we went with one manufactured by Kahrs, out of Sweden. We're very pleased with it; it was easy to install and it's proving to be a very good heat conductor.

The third option, and probably the best, is tile or decorative concrete. Not only is tile/concrete a good conductor of heat, it does not need an insulating underlayment, as is required for carpeting or wood flooring. We used tile on our bathroom floors, wood in the kitchen and great room, and carpet in the office. Every room in the house stays plenty warm, even on the coldest days, but the system does have to work a little harder to heat the office. Just the same, I enjoy the feel of a carpeted office, so—for me and the dogs at least—it's worth the small loss in efficiency.

Installing Radiant Floor Heating

There are three ways to install radiant floor heating. I have already discussed the most efficient method, wherein the PEX tubing is stapled to the subfloor, then embedded in 1½ inches of gypsum concrete (or in the case of a garage or basement floor, 4 inches or more of regular concrete). Since the heating tubes are in direct contact with a considerable thermal mass, the transference of heat from the circulating water is optimized.

If you use this method—and I highly recommend that you do—there are three further considerations to bear in mind. First, you will have to allow for an extra 1½ inches of ceiling space to accommodate the thickness of the Gypcrete®. Second, any framed walls will have to be double plated on the bottom, so

The PEX tubing is fastened to the wire mesh that lays on top of the insulation in our garage floor.

that you have a nailer for wallboard and baseboard. And third, you may have to beef up your floor specifications to support the extra weight. Usually, increasing the thickness of the floor joists by one size (from 8 inches to 10 inches, for instance) is sufficient. Your local building department will be happy to tell you what you need to do. You may find, as we did, that your floor is strong enough without modifying the design.

But what do you do if you've already built the floor and discover that it is not strong enough to support the weight of the gypsum concrete? Well, short of shoring up the floor from beneath to hold the extra weight, you can always screw (or nail) a series of plywood sleepers to the subfloor, leaving just enough room between each one to push the heat tubing snugly into the groove. Setting the tubing down into a preformed, flanged aluminum sleeve will help to direct the heat upward, where its needed.

The last—and positively the worst—method is to run the PEX tubing beneath the subfloor, through holes drilled in the floor joists. The tubing is stapled to the underside of the subfloor, and then covered with aluminum heat transfer plates to reflect the heat into the subfloor. Insulation should then be installed beneath the heat tubing to further prevent heat loss. The shortcomings of this system are obvious, since all the heat must pass—and therefore, disperse—

Gypcrete® is poured over radiant heat tubes on our main floor, and leveled to the height of 2 x 4 sleepers.

through the subfloor before it can even be felt on the floor above. For folks on a strict energy budget, this may not be acceptable.

■ Incorporating Solar-Heated Water Into Your Heating System

While a good passive solar home design is the very best way to use solar energy to heat your house, a hot water heating system can be readily adapted to use free energy from the sun to actively augment a passive solar home's efficiency.

Typically, water (or, in colder climes, a freeze-proof glycol solution) is heated by the sun in a series of solar collectors—weatherproof, glass-covered panels through which snake loops of black copper tubing. Though usually mounted on the roof, the solar collectors can be located anywhere near the house, as long as they have unimpeded access to direct sunlight, and a path to run tubing into the house.

There are many ways to set up a solar-heated water system. Typically, the water heated in the solar collectors is circulated—via an electric circulating pump—through a heat exchanger located near a large water storage tank. Water for home heating is then warmed by the heat exchanger as it circulates through the storage tank, thus pre-heating it before it reaches the boiler or water heater.

The efficiency of such a system is dependent on many factors. The most important, the daily hours of solar radiation, is probably the

Five solar hot water collectors are installed next to smaller photovoltaic panels on Curt Busby's home.

least of your concerns if you're using sunlight for your electrical needs; if you have enough sun to drive your PV array, you should have enough to heat your water.

As with your PV array, the angle of the solar collectors in relation to the sun is important. Since it is highly unlikely that you will make seasonal adjustments of the hot water solar collectors, set the angle for winter, when you will need most of your hot water. (An angle of latitude +15 degrees, you will remember, is ideal for capturing winter sunlight. If this isn't feasible, they should, at the very least, be within 15 degrees of latitude, one way or the other, and, of course, facing as close to south as you can get them.)

The surface area of the solar collectors is important. The bigger they are, the more water they can heat—that much is intuitive. But the solar collectors will also need to be sized for the tank and the heat exchanger. If the tank is too big, or the collectors too small (or too few), the water may not heat up sufficiently to do you much good.

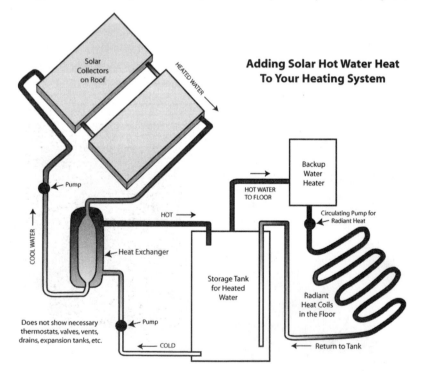

Adding Solar Hot Water Heat To Your Heating System

A Better Solar Hot Water Collector?

Flat-plate solar collectors have been around for a long time. We all recognize them as the heavy, clunky, dim-looking rectangular panels that adorn so many residential and commercial roofs (often commercial laundries). But they aren't the only technology out there—evacuated-tube solar collectors are proving to be a viable alternative to flat plate collectors.

Evacuated-Tube Solar Collectors

What's the difference? Flat-plate collectors simply run water through a serpentine loop of copper pipe imbedded in an insulating material within a metal box covered with tempered glass. It's a simple technology, which is good, since simple systems are usually more reliable than complicated ones. But flat-plate collectors do have a few drawbacks. For one thing, they're exceedingly heavy, even when they aren't filled with water. But the real Achilles heel lies in the fact that they can only operate really efficiently when the sun is directly overhead, since that's the only time the sunlight will be perpendicular to the collector.

Evacuated tubes take a different approach. Consisting of rows of cylindrical glass tubes that are exposed to direct sunlight at practically every angle, they capture more heat. The collector is a coated inner glass tube enclosed within an outer glass tube and separated by a vacuum. Water-filled copper channels run within the inner tube, which absorbs solar energy; energy which cannot flow back out through the vacuum. The result is hotter water in a smaller volume. Evacuated-tube systems are lighter and more efficient than flat-plate collectors, so they take up less space and are far easier to install. The downside? They cost more. Are they worth it? In northern climes with diffuse sunlight, certainly. And for those of you with limited roof space, they just might be the answer.

Mazdon evacuated-tube solar collectors installed on University of Colorado's 2005 Solar Decathlon home. One array is installed on the roof and one on the wall.

On the other hand, if the tank is too small in relation to the solar heating capacity of your collectors, you will end up wasting valuable solar radiation because the water in the tank will be quickly heated to capacity during the day, but will soon lose much of its heat at night.

The best thing you can do before spending a lot of money on an improperly-sized system is to seek advice from someone in your area who routinely installs solar hot water heating systems. Even if they charge you a consulting fee, you'll still be money ahead in the long run.

Wood Stoves

A fireplace is a wonderful, romantic setting for relaxing on cold winter nights, and if you plan to put one in your new home, I think it's great. There's nothing like the sight of jumping flames to excite the imagination and soothe the spirit.

But if it's heat you want, you will be much better off with a centrally located, wood-burning stove. It might not be as aesthetically pleasing as a fireplace, but it may save an argument or two over the cost of heating your home.

Although it's an ancient technology, burning wood for heat is still sensible and cost effective. Wood, like wind and sunshine, is a non-depleting source of power, since most firewood is standing dead, or culled from overgrown forests that need to be thinned. And, with the efficient stoves now on the market, wood is a much cleaner fuel than ever before.

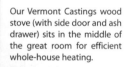

Our Vermont Castings wood stove (with side door and ash drawer) sits in the middle of the great room for efficient whole-house heating.

The greatest thing about using a wood stove in an off-the-grid home setting is the availability of fuel. Unless your home is in a most unusual place, you should have plenty of wood on, or near, your property.

Even if you have to buy it, it'll almost certainly be cheaper than natural gas, propane or heating oil.

If someone offers you a great deal on an old, pre-1988 wood stove, you should respectfully decline the offer. Why? Because 1988 was the year that wood stoves entered the modern world. Concerned about the growing problem of air pollution and wood stoves' contribution to it, the EPA sat down with wood stove manufacturers and kindly asked (as only a government agency can) that all new wood stoves be designed to meet strict emission standards. The result was a pair of new designs that dramatically reduced emissions and greatly increased efficiency in the bargain.

One of the improved designs utilizes a catalytic converter, similar to the one in your car or pickup, that enables the stove to burn compounds within the smoke that would normally go up the flue and into the atmosphere, unburned. The extra combustion means cleaner air, more heat with less wood, and a stove that can hold a fire longer than any of its predecessors.

Another EPA-approved design that accomplishes pretty much the same thing as a catalytic stove without the converter, simply circulates the gases back through the stove to be burned a second time, before exiting up the stove pipe. Called "secondary combustion" stoves, these stoves boast the advantage of eliminating a costly element—namely, the catalytic converter—that will have to be periodically replaced.

Which kind of stove should you buy? Specifications will vary from one manufacturer to the next, but as a general rule catalytic stoves are more efficient, cleaner and more expensive than secondary combustion stoves. Our cast-iron, catalytic stove easily holds a fire all night long, even with fast-burning pine in the fire box. The key is to keep the stove pipe free of creosote buildup—we clean ours every couple of months, just to be sure—and the doors adjusted so they close tightly.

Will a wood stove save you money? At $80 per cord, $200 worth of wood burned in the stove will save us around $600 worth of $2-per-gallon-propane burned in the boiler. Besides, chopping wood is much better exercise than writing out checks to the propane company.

Masonry Stoves: Heat Batteries

Ask anyone what the purpose of a battery is and they'll tell you it's to store electrical energy for future use. And that's the idea behind a masonry stove: to store up heat to be given off slowly over time.

Masonry stoves have been around for hundreds of years. They were originally built as a means to conserve firewood in Northern Europe, primarily Scandinavia. The idea behind them is simple: to store as much heat energy as possible from a given amount of wood.

Conventional wood stoves begin to give off heat the minute the fire is lit. They heat up quickly, and if you're cold it's a nice feeling. In short order you'll find yourself closing the draft and shutting down the damper to keep the room from getting too warm. But by shutting off the stove's air supply you are reducing the combustion efficiency of the burning wood. And even with the stove shut down, most of the heat goes up the chimney and is lost to the great outdoors, since there is not much between the fire and the stovepipe to stop it.

A Tulikivi stove made of soapstone is one style of masonry stove. Many are custom-built from local stone. *Photo courtesy of Tulikivi.*

Masonry stoves are just the opposite. They don't rely on the heat-conductive properties of metal to quickly transfer heat to the room; instead, they use the thermal mass of brick and stone to soak up the heat and give it back slowly. They do this by routing the flue gases through a long series of baffles built into the structure; baffles that keep the gases moving around and giving up heat every step of the way.

Generally, owners of masonry stoves will burn a single hot fire every one to three days. Once the stove's thermal mass heats up it takes relatively little wood to keep it warm. The heat given off varies little from one hot fire to the next, since the stove's tremendous mass acts as a constant heat radiator.

Masonry stoves should be built as close to the center of the house as possible, and will perform best in homes with lots of thermal mass to soak up the constant heat. These include log homes, rammed-earth homes, adobe homes, and straw-bale homes—all the really cool homes we're secretly looking for an excuse to build. One drawback is the square footage required for this type of heater, but the benefits should easily offset the loss in room space.

Wood-Fired Boilers

For those of you wishing to lower or eliminate your natural gas, propane, or fuel oil bill, a wood-fired boiler might be the answer. These stout units operate in much the same way as a regular boiler, in that they use the heat from burning fuel to heat water running through a heat exchanger. This hot water can then be used for domestic purposes or to heat the home. And since they are designed to burn wood with a very high temperature of combustion (up to 1,800 degrees Fahrenheit; a little above to the melting point of silver), a wood-fired boiler is considerably more efficient than a regular wood stove.

To get the most out of one of these units, you should combine it with an insulated hot-water storage tank of several hundred gallons. In this way the excess heat produced above and beyond your immediate needs can be stored for future use, much the way heat is stored within the mass of a masonry stove. As an extra bonus to using a storage tank, the boiler can be made to operate in tandem with a solar hot-water system.

Though they only need to be stoked once every day or two, it can be an inconvenience for those of you who can't always be home that often. For you the answer might be a multi-fuel boiler, one that can automatically switch over to propane, natural gas or heating oil when the temperature in the combustion chamber falls to a preset level. Then, when you get home, you simply light another fire and the unit switches back to wood-heat mode.

If you would prefer to stick to a single fuel (and who wouldn't?), HS-Tarm, a Danish company that markets its products in the USA through Tarm USA Inc., produces a wood-pellet-fired boiler that automatically feeds itself, much like a pellet stove *(see below)*. All you need to do is fill a bin with pellets and let the boiler take care of itself. (In many places in Europe, where biomass heating is rapidly gaining in popularity, you can phone in a delivery of wood pellets right to your bin, just like you order propane.)

A word of caution: since a fan is used to force air into the combustion chamber, these units do require some electricity beyond that used to run the circulating pumps for the hot-water heating system, so check with the manufacturer before you buy and then plan accordingly. Additionally, if you are considering a pellet-fired boiler, extra electricity will be needed to run the auger. To learn more about wood-fired boilers, check out *www.woodboilers.com*.

Corn and Pellet Stoves

Note: Corn and pellet stoves use electricity to run the auger and blower motors. How much? To answer that question, we loaned our Watts Up? meter to Golden Grain Corn Stoves. Set to burn one bushel per day (an average setting), the stove used around 159 watts per hour, or 3.82 kWhs over 24 hours. This exceeds the energy usage of three efficient refrigerators, making these stoves highly impractical for off-the-grid homes.

As a former breeder of race horses, I can attest to the potency of corn as a high-energy fuel (if you have any doubts, read the calorie content for a handful of corn chips). The main advantage of corn—aside from the fact that it is such a clean fuel—is its availability. You can buy corn anywhere people keep livestock; just about everywhere, in other words.

Corn stove by Golden Grain Corn Stoves.
(photo courtesy of Golden Grain Corn Stoves)

Corn stoves differ from wood stoves in that they have very small combustion chambers and do not require the expensive chimney systems needed by wood stoves; the small amount of exhaust gas is much cleaner—and cooler—than wood smoke and can be safely vented horizontally through a wall.

You should call a few feedmills to ascertain the local cost of whole corn (14 – 15 percent moisture is ideal) before investing in a corn stove. While almost all of the comparisons I found for the cost-effectiveness of corn stoves listed corn at $2.50 per bushel, the cheapest corn I could find locally was $5.90 per hundredweight, or $3.30 per (56 lb.) bushel. This was in Weld County, Colorado, no less, where cornfields are even more ubiquitous than tacky housing developments.

Pellet stoves work on the same principle as corn stoves. But instead of corn they burn pellets made from lumber mill scraps, agricultural refuse, or even waste paper and cardboard. American Energy Systems makes a multi-fuel stove that burns corn, pellets, and even cherry and olive pits. Most of these things would have been plowed under or left to rot in bygone days, but now they are considered biomass, a broad classification of environmentally friendly, plant-derived fuels that are both renewable and carbon neutral.

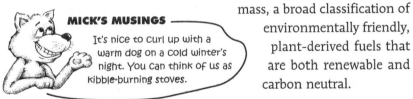

MICK'S MUSINGS

It's nice to curl up with a warm dog on a cold winter's night. You can think of us as kibble-burning stoves.

Whence Flows My Domestic Hot Water?

Realistically speaking, you have three choices for your domestic hot water supply, any one of which can be incorporated into a solar heated water system. The most common of these is the "good-old" glass-lined tank with a gas burner on the bottom. They're cheap and use no electricity, which is good, but they're all bound to fail in a few years, which is bad. You also end up heating a lot of water that's going to cool off and have to be reheated before you get around to using it, which makes it wasteful.

An indirect water heater is another option. These units are supplied with water from the home heating system (the boiler), by way of a heat exchanger. On the plus side, they last far longer than conventional water heaters and, though the initial cost is greater, they will save you money over the long haul. On the minus side, they have the same problem of leaking heat while in standby mode, and they also require an electric pump to move water through the heat exchanger.

Tankless on-demand water heaters are your final and best choice. These units heat water in a compact gas-fired burner as you use it. Formerly suited to nothing grander than a weekend cabin, on-demand heaters have gained a lot of sophistication and well-earned respect in recent years. Although the larger models (5-plus gallons per minute) are expensive, they last practically forever and can pay for themselves in a few short years on the energy savings alone. Before you buy, it's important to know that some models will gladly accept pre-warmed water from a solar collection system, while others won't. Be sure of what you're getting.

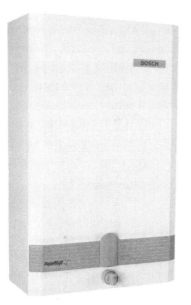

An excellent choice for off-grid homes is the Bosch AquaStar 125HX—a pilotless, on-demand heater that doesn't require electricity.
Photo courtesy of Controlled Energy Corporation.

Curt and Kelly's Excellent Solution

One of the nicest things about living where we do is our proximity to like-minded people who have engineered their own solutions to producing and saving energy. Curt and Kelly, our nearest neighbors, have come up with a clever and cost-effective solution for heating both their 1,300 square foot house, and their domestic hot water.

To begin with, they use five 4-foot x 10-foot roof-mounted solar collectors to heat the water (*see photo on page 229*). Curt managed to pick up this impressive array from a classified ad, for a fraction of its original cost. The seller needed to re-roof his house and had decided not to re-install the collectors after the job was complete. *continued*

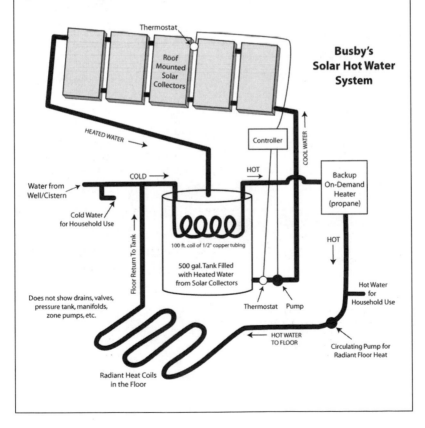

Busby's Solar Hot Water System

Thermostat

Roof Mounted Solar Collectors

HEATED WATER →

Controller

COOL WATER →

COLD →

HOT →

Backup On-Demand Heater (propane)

Water from Well/Cistern

Cold Water for Household Use

Floor Return To Tank

100 ft. coil of 1/2" copper tubing

500 gal. Tank Filled with Heated Water from Solar Collectors

HOT

Does not show drains, valves, pressure tank, manifolds, zone pumps, etc.

Thermostat   Pump

Hot Water for Household Use

← HOT WATER TO FLOOR

Circulating Pump for Radiant Floor Heat

Radiant Heat Coils in the Floor

To provide storage capacity for such a bountiful harvest of hot water, Curt and Kelly installed—which is to say, built their house around—a highly-insulated 500-gallon rectangular steel tank. Water from the tank circulates through the solar collectors and back, gradually raising the temperature inside the tank as the day progresses. On most sunny days they are able to bring the water to—and hold it at—a temperature near boiling. A differential temperature thermostat hooked to a pair of temperature sensors (one at the tank, another between two solar collectors) activates a pump that circulates the water through the system as it's heated by the sun. Whenever the collectors are cooler than the water in the tank, the pump shuts off, causing the water to drain back down into the tank. This is a handy feature to keep the water from freezing inside the copper loops within the solar collectors at night, and on cold, cloudy days. It also eliminates the need for antifreeze (glycol).

A 100-foot coil of copper tubing running through the tank serves as a heat exchanger for both the radiant floor heat, and the domestic hot water. Being so massive (and efficient), Curt and Kelly's system does not require an expensive boiler, or even a standard hot water heater, to supply additional hot water during off-hours (or off days). Instead, the heat-exchanging coil runs from the tank through a propane-fired, on-demand hot water heater. It's all they need to heat their home and domestic water.

## chapter 13

Pumping Water

getting it from the ground to your house,
without overtaxing the system

This chapter is included for the many readers who have undertaken the task of learning about renewable energy for the purpose of using it to power a remote home. For those of you in a more urban setting, the information presented here may be nothing more than a curiosity. Just the same, there's lots of good stuff here on which to exercise your imagination.

Water is your property's most valuable asset. Electricity you can make; propane can be brought in by truck. But if your land doesn't have sufficient potable water, your life will

One of many old-fashioned wind generators in windy Wyoming that was, and still is, used to pump water for cattle.

revolve, to a large extent, around the transportation and storage of this life-giving liquid.

In some places ground water is abundant; in others it's hit-or-miss affair. Our neighbor to the west got a good well (3 gallons per minute) at 340 feet, while our neighbor to the east got a trickle (5 gallons per *hour*) at 700 feet. We had no idea what to expect, but after watching the driller sink a hole 480 feet through impermeable rock with no water in sight, any optimism we earlier felt quickly dissolved. Then, like magic, the morphology of the rock changed and water appeared. Lots of it. By the time the drill bit reached 540 feet, we had a well producing 5 gallons per minute. We let out the breath we'd been holding for several days, and uncorked a fine Merlot.

Our problems were far from over, of course. Being off the grid and therefore on a strict energy diet, we still had to figure out the best way to get the water from the bottom of the well to the house, but at least we were dealing with definable parameters. After what we'd just been through, it seemed like a manageable concern.

MICK'S MUSINGS

I've observed that cats don't care much for water. It must dilute the caustic slush that flows in their veins.

Just the same, we carefully weighed the options—numerous and varied as they were—before deciding what we'd do.

Should You Install a Cistern?

Before buying a pump, you will need to decide if you are going to pump water into a cistern and then pressurize the house water with a much smaller pump, or simply forget the cistern idea and pressurize the house directly from the well pump. There are three primary reasons people use cisterns: low producing wells, not enough energy to run the well pump, and water for fire protection.

◼ Low Producing Wells

People who have wells with very low recharge rates use cisterns to provide a buffer between what the well can store within its casing,

and the amount of water they might
need to use within a short period of
time. As an example, let's say that you
have a well with a paltry 5 gallons per
hour recharge rate. In one day, it will
provide 120 gallons of water. Not

Reasons for a Cistern
■ Low producing wells
■ Not enough energy to run the deep well-pump
■ Water for fire protection

much, but enough for two people, aware of the limitations. But if the
well casing only holds 70 or 80 gallons, that's all that can be used in
a short period of time. However, by pumping the entire contents of
the well casing into a 1,000 or 1,500-gallon cistern every time the
well is fully recharged, you will be assured of always having enough
water, even though the well is a poor one. A really deluxe setup uses
a float system within the cistern (similar to the one in a standard
toilet) to automatically turn the well pump on and off (and also to
start a generator). A sensor within the well can shut down the pump
when the water level falls too low.

■ Not Enough Energy To Run The Well Pump

There seems to be almost a paranoia about running an AC deep-well
pump with a PV/wind system. This is partly because many people
don't want to commit that much precious wattage to running a high
amperage pump, and partly because they don't want to push the limitations of their inverters. Whether their fears are well-founded or
not, most people on PV/wind systems with deep well pumps choose
to pump their water into a cistern with a fossil fuel-fired generator or
a stand-alone, direct solar-powered DC submersible pump. They then
pressurize their house water line with a small AC (or DC) pump. I'll
be the first to admit that these people are right, on both counts: it *does*
take a fair bit of energy to run a deep-well submersible pump from
the batteries, but not all *that* much; and it does tax the inverter at
times, though (in our case, at least) never enough to threaten the
integrity of the system. We knew we were asking a lot from our
PV/wind system by using it to pump water from 540 feet down with
an 11-amp, 240-volt pump, but we had to try it, even though many
people told us it wouldn't work. The reason we did it is simple enough:

Rainwater Collection

More and more people are collecting rainwater these days. We use our roof rainfall to keep our fire-protection ponds topped off (and the deer watered), and to fill barrels that water LaVonne's irises and gladiolus (which the deer snack on after drinking from the pond). Just how much water will you collect from your roof? I've come up with a painless formula for determining how much water falls over a given area:

Area of roof (sq. ft.) x inches of rain x .623

(.623 is derived from 0.0833, the number of feet in an inch,
times 7.48, the number of gallons in a cubic foot)

If ½-inch of rain falls on a 1,200 square foot roof,
how many gallons are collected? **1200 x .5 x .623 = 373.8 gallons**

A lot of water! You may need more barrels than you thought.

by automatically pressurizing the house water with a deep well pump, it's one less thing to think about, meaning that we never have to worry that we'll be soaped-up in the shower some morning, only to discover that one of us (namely, me) forgot to fill the cistern. Besides, we don't even like to listen to our neighbor's generator run every night from across the canyon; we like to listen to our own even less.

But don't despair; as you will soon discover, a good DC pumping system can bypass all of the above concerns, though it will open yet another can of worms. (Is there no end to the worms in this business?)

■ Water For Fire Protection

After three years of relentless drought, fire is on everyone's mind around here. Having been on the fringe of a massive fire two years ago, and within a mile of four lesser lightning-started fires last year, we think about fire a lot. The local volunteer fire department recommends that everyone have at least 2,500 gallons of water stored for fire protection. A large cistern can accommodate a good part of that amount. Since we don't have a cistern, we placed a 1,500-gallon agricultural tank outside the house, then built a pair of ponds to hold an additional 1,000 gallons. We figure after the sheriff forces us to evacuate (if he can find our home), the firefighters will have a much easier time finding a pair of ponds and a big, green, above-ground tank than they will an indoor, or below-ground, cistern that will more than likely *not* be full when they need it.

Well Pumps

There are two broad categories of submersible well pumps: AC and DC. Both have their strong and weak points, and neither can be used successfully in every application. In most instances, I agree with Demetri, the venerated pump installer who services most of the wells in these hills, when he says that you should never ask a DC pump to do what an AC pump can do better. But I'm getting ahead of myself.

■ DC Well Pumps

DC-operated well pumps can withstand a range of voltage that would quickly destroy an AC pump. Because of this, they are used primarily in stand-alone systems. This means that you can wire them to their own solar array, or wind turbine, and forget about them. They will pump water whenever the sun shines, or the wind blows. They are designed to work with whatever power is available to them (within limits, of course), so long as it's enough to start the pump turning. Many models can be run dry without sustaining damage, which makes them quite useful in wells with low recharge rates. With the addition of a pump controller, the pump can be wired into a float switch that shuts off the power when the cistern is full, and starts it again when the water drops below a preset level.

Admittedly, it's an attractive idea. With a big enough cistern, you'll never have to worry about having enough water, even during a cabin-fever-inducing run of non-productive weather. And, by having the pump hooked to its own power supply, you won't have to draw from your household energy savings—or drag out the generator—to shower or wash the dishes.

So, how much is it worth to you to never have to worry about water? (That's always the rub, isn't it?) A good DC pump is an expensive proposition. And the wire doesn't come cheap either. For example, a 48-volt DC pump, rated at a mere 20 percent the capacity of our 240-volt AC pump, costs twice as much, and requires wire 6 gauges heavier. And, on average, we figure that our water pumping takes about 200 watts of our 1,140-watt array. To power a DC pump rated for a well the depth of ours, we would have to dedicate an additional 120 watts (320 watts total) of solar power.

But if money were the only problem, we still might have gone with an install-it-and-forget-about-it DC pumping system. Except it's not exactly that easy. Most DC motors have brushes that have to be periodically replaced. And, we were told, at our well depth we should expect certain pump components to fail earlier than normal. The thought of pulling up 540 feet of drop pipe at the whim of a fractious pump motor was enough to drive us firmly into the AC camp.

Had our well been shallower, say 200 feet, the pendulum might have swung the other way.

DC Well Pumps

Conergy (formerly Dankoff) Solar pumps are ideal for stand-alone DC pumping systems. Not only are they efficient, they can deliver water from depths up to 650 feet. Check them out at *www.conergy.com*.

■ AC Well Pumps

In the race to supply your water, DC pumps are the tortoises: they plod along slowly and ceaselessly. That means, of course, that AC pumps are the hares—on steroids. They're lean and mean and tough. They consume a lot, but they produce even more. They can pump from virtually any depth the well driller can find water, and you can expect them to last for many, many years without service or replacement.

Since AC pumps move such a high volume of water (our 1.5 hp pump delivers 6 gpm, even at 540 feet) the total *amount* of energy they use is surprisingly small. As mentioned above, we figure, on average, 200 extra watts of solar capacity is enough to power our well pump. The problem is, an AC pump needs so much power *all at one time*. Specifically, we're talking about the amperage the pump demands in the split second when it goes from "off" to "on", which may be as much as three times higher than its rated amps.

For us, as I've said many times in these pages, it's not really a problem. But it may be for others. Our well pump operates nicely within the rated capacities of our Trace SW4024 inverter, as long as we're mindful of other loads that might be operating at the same time. To be running the clothes washer and gas-powered dryer, a stereo and a couple of computers the instant the pump kicks-in does not present any difficulty. But, if we were also running the dishwasher, we might be pushing the system toward the edge. And throwing a table saw into the equation would certainly cast it into the abyss. Viola! Our own personal blackout. (But really; how much stuff do you need to run at one time, anyway?)

A larger inverter, or two stacked inverters would alleviate the problem, and if you go with a watt-gobbling AC well pump, your PV/wind equipment supplier will almost certainly try to convince you to

buy two inverters. The choice is yours. If it turns out you have difficulty running a large well pump with a single inverter, the problem may lie more with the pump, than the inverter. This is because all AC well pumps are not created equal; some require more power to start than others *(see inset)*. Your pump installer should be made aware of the limitations of your inverter(s), and should be able to sell you a pump that falls within the parameters set by the inverter manufacturer.

Specifically, you will want a pump that requires a separate starting box outside the well (inside the mechanical room), rather than a pump that has the starting circuitry built into the motor casing. A simple relay-type starting box will work better with an inverter than an electronic one.

Also, if you use a 240-volt transformer to supply power to the well pump, it is less work for the inverter if you place the transformer between the pressure switch and the starting box, rather than between the inverter and the pressure switch.

Explain these things to your pump installer. He or she should know exactly what your concerns are and how to remedy any potential problems. If not, there are always other installers down the road.

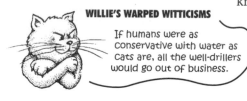

WILLIE'S WARPED WITTICISMS

If humans were as conservative with water as cats are, all the well-drillers would go out of business.

Decisions, Decisions

It's not easy to weigh all the pros and cons of all the different ways people have conceived to deliver water from the earth under your feet to the sink in your kitchen. Just when you think you have it all figured out, some intractable fact, lurking in the shadows, jumps out and trips you. Happens to me all the time.

Odd as it may seem, the best remedy for too many facts is more facts. Talk to people on every side of the issue, AC and DC. Talk to your neighbors. Talk to your well driller and your pump installer. Call the inverter manufacturer. Tell them all what your particular circumstances are.

Then follow your instincts. You can't go wrong that way.

Personal Power Companies

Homeowners	Jerry & Lois
# of Occupants	4 adults In 2 homes
Location	Heber, Arizona
Home Size	4,000 square feet on one level and small loft, plus a mobile home and shop
Home Heating	Quadra-Fire wood stove (primary source) and propane-fired forced air (rarely used) in main house
Water Heating	Conventional propane-fired water heater
Grid-Tied?	No
PV Array	5,000 watts +, in 6 ground-mounted arrays south of house
Charge Controller(s)	5 Xantrex C40s
Inverter(s)	Xantrex SW4024 sine wave
Batteries	32 Rolls Surrette flooded lead-acid; 3,000 amp hours @ 24 volts
Wind Turbine /Tower	400-watt Southwest Windpower Air X turbine on a 20-foot pipe tower
Solar Hot Water	None
Backup Generator	15 kW diesel-fired, wired for automatic start
Comments	This remarkable restored 19th century homestead is powered by a single inverter which is monitored and controlled from Jerry's home office via a computer link. A bunkhouse runs from a separate 12-volt system. Water is pumped by the generator from a 600-foot well into a pair of large holding tanks and gravity fed into the house. The entire installed system cost around $45,000.

## chapter 14

Final Thoughts

learning to trust Mother Nature

So, here you have it. Most of the knowledge I've gained through hard-won experience is in this book, as well as a great deal of what I've learned simply because my curiosity would permit me no rest if I hadn't. But for all that, this book is still just an introduction to the burgeoning subject of renewable energy. Even if you've read every page, you still have a great deal to learn—we all do. Much of it you'll find in articles and books far more technical than this one, or scattered about in a million nooks and crannies in the nebulous maze of cyberspace. But the real pearls of knowledge—those that bury themselves so deeply in your brain that they can never be dislodged—will be those you discover for yourself.

If, as I hope, I was able to convey in these pages the enthusiasm I feel for the prospects of solar, wind and micro-hydro power, it is because my enthusiasm is real; real because these non-depleting sources of energy have for me become not only a means to an end, but a way of life. I cannot imagine waking up in the morning without parting the drapes and studying the sky, or watching the treetops to see how hard the wind is blowing, and from what direction. Or feeling a tickling urge to step outside to feel the brisk morning air against my face, so I can get a sense for what Mother Nature is serving up today.

To live in step with the rhythms of nature, I have learned over the years, is more than just sharing in the splendor of every bright, sunny day. It is also to acquiesce to being humbled by nature; to realize you are without redress, and to accept whatever comes your way, even if it seems—as it often does—that you are being treated unfairly. The sun shines where it will. The wind blows as it chooses, but no one knows when, or where, or how strong—or for how long. And still you have to believe that the fickle sky will provide for you.

That's the beauty and the paradox of this whole enterprise.

Once you take the plunge and install your own system—whether you do it now, of your own free will, or you wait until the squirrels drive you to it—you'll discover exactly what I'm talking about.

Until then, I'll leave you with a little metaphor to ponder...

Changing with the Weather

A METAPHOR

Imagine that all of your fresh water comes from a small, clear spring. You know, instinctively, that if you use the water wisely, there will always be enough. Right away, you notice that when the sun shines, and the wind blows, the water flows. Sometimes it gushes; other times it merely trickles. On calm nights, or when the clouds roll in so thick the wind can't blow, the flow of water stops altogether.

All day long, every day, you fill vessels from the spring. When they overflow with water you bathe and wash and clean and drink deeply. But when darkness comes, and the skies grow still, you take only what you absolutely need.

At first, it seems a capricious existence. After all, you used to live next to a bottomless river; no matter how much water you removed, the river seemed just as full. But life is vastly different now, and you can't go back to the way it was before.

You have always thought that weather was a fickle, random thing, but soon you begin to discover certain underlying patterns you've never noticed before. You see that the breeze blows one way during the day, a different way at night, and storms are almost always heralded by winds. The brightest, clearest days are those right after a storm system passes. An abrupt change in temperature means an increase in the strength of the wind. Each season, you discover, comes with a different breed of clouds, and its own special wind.

Every morning you watch where the sun rises, and where it sets in the evening. You watch to see how high it climbs at midday. You come to know how much water will flow at each hour of the day, in every kind of wind, in every degree of cloudiness.

Each day that you learn something new, your fear of running out of water grows less acute, for you have come to know—almost as a matter of instinct—when your vessels will overflow, and when they will run nearly dry.

After a time, as you begin to learn the rhythms of nature, your initial annoyance at being occasionally inconvenienced evolves into a feeling that approaches reverence for the ever-changing skies. The heartbeat of nature becomes the heartbeat of your house, and of your life. You become attuned to your surroundings in a way you never thought possible before. Though the weather is still unpredictable, for the first time in your life you begin to trust in constant change, for you know that the water you were denied one day will be replenished the next. And, even though you are using far less water than ever before, somehow you don't miss the excess.

Finally, on the day when you truly understand that whatever nature gives you is exactly what you need, a new, unfamiliar feeling will wash over you.

Don't worry—it's called Freedom.

Acknowledgments

thanks!

Writing a book is rather like building a house: to get the job done you have to lie to yourself about how easy it's going to be. Otherwise, the house would remain so many crude lines on a Big Chief tablet and the book would forever languish in vaporous dreams of menacing grandeur. Self-deception alone is not enough, however. A book such as this also requires the talents of far more people than the glory-hog whose name appears on the cover. To my good fortune, I've made the acquaintance of quite a number resourceful, experienced, imaginative, and downright smart people (and critters). And now it's time to thank to them all.

First I must thank my wife, LaVonne, for her easy-to-grasp technical illustrations, enticing sense of design, sharp editorial skills and, of course, her diplomatic dispensation of encouragement and prodding.

For her hilarious illustrations—especially those that brought the fable to life—I especially want to thank our good friend, Sara Tuttle.

For further editing and proof-reading my book, while asking only that I critique her novel in return, I thank author and friend from across the great snowy meadow, Linda Masterson.

Doug Pratt, former Real Goods' technical editor, embraced *Power With Nature* as though it were his own, and made it inestimably better for his efforts. Thanks, Doug!

Thanks also to the late Mark "Dr. PV" McCray of RMS Electric for sharing so much what he learned from installing hundreds of systems over the years. You are greatly missed, Mark. To Dr. Ronal Larson of the American Solar Energy Society, for casting his keen, technical eye across my manuscript and then making many valuable suggestions. And to Sam Burnham, of Burnham Beck & Sun, for proofing the grid-tie chapter for accuracy.

For helping me to understand what really goes on inside those mysterious humming components in my electrical room, I thank Marty Spence, Robin Gudgel and Christopher Freitas from OutBack Power Systems.

Many fine companies provided photos and information, and I thank them one and all, just as I thank all of the friends and neighbors whose houses, solar arrays and solar collectors appear in these pages.

And finally, I fondly wish to thank my very best of friends, Big Mick, Wild Willie, Newt, Amy and Stinky for providing such lively inspiration for the fable. You are all more real—and more fun—than mere words could ever convey.

APPENDIX A

Checklist of Equipment

Solar Electric System
solar modules
frame for PV array

Wind System
turbine & charge controller
tower

Micro-Hydro System
penstock
turbine-generator
regulator & diversion load

Components
inverter(s)
charge controller(s)
DC disconnect(s) & breakers
120/240 transformer
metering

Batteries
batteries
temperature sensor
cables
battery box
venting

Wiring
from solar, wind & hydro
in between components
for grounding
conduit & fittings

Misc.
combiner boxes
junction boxes
breaker boxes & breakers
ground rods & clamps
terminal connectors
colored tape

Solar Hot Water
solar collectors
piping & fittings
tank & heat exchanger
circulating pump(s)
temperature sensors/thermostat

Compact Fluorescent Bulbs

Lightning Protection

Special Appliances
gas range (without glow bar)*
new Energy-Star refrigerator
front-loading clothes washer
gas clothes dryer

Peerless Ranges do not need electricity, and they do not have a glow bar for the oven.

Other Costs To Consider
erecting wind tower
installation of components
  and wiring
shipping/freight
any consulting fees

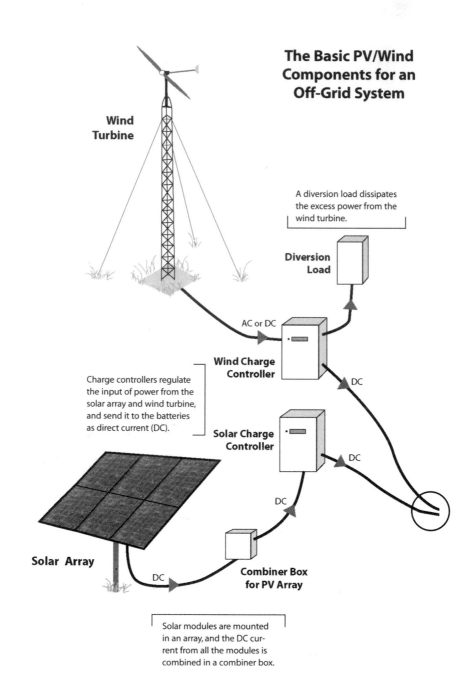

The Basic PV/Wind Components for an Off-Grid System

Wind Turbine

A diversion load dissipates the excess power from the wind turbine.

Diversion Load

AC or DC

Wind Charge Controller

DC

Charge controllers regulate the input of power from the solar array and wind turbine, and send it to the batteries as direct current (DC).

Solar Charge Controller

DC

DC

Solar Array

DC

Combiner Box for PV Array

Solar modules are mounted in an array, and the DC current from all the modules is combined in a combiner box.

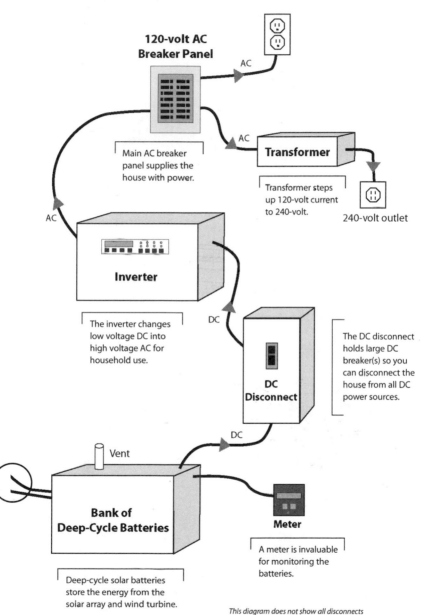

120-volt AC Breaker Panel

AC

Main AC breaker panel supplies the house with power.

AC

Transformer

Transformer steps up 120-volt current to 240-volt.

240-volt outlet

AC

Inverter

The inverter changes low voltage DC into high voltage AC for household use.

DC

DC Disconnect

The DC disconnect holds large DC breaker(s) so you can disconnect the house from all DC power sources.

DC

Vent

Bank of Deep-Cycle Batteries

Meter

A meter is invaluable for monitoring the batteries.

Deep-cycle solar batteries store the energy from the solar array and wind turbine.

This diagram does not show all disconnects and fuses, wind brake, grounding, etc.

Power Consumption of Appliances
We Measured with a WATTS-UP? Meter

Appliance	Continuous Draw (Watts)
Computer, desktop	90
Computer, laptop	24
17" monitor	100
17" LCD (flat screen) monitor	50
HP LaserJet printer (in use)	600
HP Inkjet printer (in use)	15
Microwave	1,400
Coffee Maker	900
Toaster, 2-slice	750
Amana Range (propane): Burners	0
Oven (with glow bar; when heating)	380
Blender	350
Mixer	120
Slow Cooker (high/low)	240 / 180
20" Television	50
27" Television	120
27" LCD Television	95
50" LCD Television	190
50" Plasma Television	330
Stereo System	25
Stereo, portable	10
Vacuum, Oreck	410
Vacuum, Dirt Devil Upright	980
Table top fountain	5
Sewing Machine (Bernina)	70
Serger (Pfaff)	140
Clothes Dryer (propane)	300
Clothes Iron	1200
Hair Curling Iron	55
Hair Dryer (high/low)	1,500 / 400
Furnace Fan (1/3 hp / 1/2 hp)	700 / 875
Corn Stove	159
Treadmill (walking/running)	150 / 800
Nautilus Elliptical Trainer	4

One watt delivered for
one hour = **one watt-hour**

1,000 watt-hours = one **kWh**

amps x volts = watts
2 amps x 120 volts = 240 watts

*We have not listed refrigerators or freezers since their efficiency is getting better every year. Look at the **EnergyStar.gov** website for the latest ratings.*

Appliance	Watt Hours
Dishwasher, cool dry	736 watt-hours/load
Clothes Washer (front-loading)	145 watt-hours/load

Notes on Appliances

Use compact fluorescent light bulbs...they add up to big energy savings. For example: if 6 bulbs are on for 5 hours a day: 60-watt incandescent bulbs will use 1,800 watt hours per day; 13-watt compact fluorescent bulbs will use only 390 watt hours.

Low-usage, high energy appliances (hair dryers, microwaves, coffee makers, etc.) are not much of a problem since they draw very little power when averaged out over time. You can also choose not to use them, if you're low on power.

Invest in a new **refrigerator and/or freezer**. You'll be amazed at how much more energy-efficient they are. The typical new fridge now uses 80 percent less energy than models from the late 1980s and early 1990s. Do your research on *www.energystar.gov* before buying and always read those yellow tags!

For a cooking range, buy one without a glow bar, which can use 300-400 watts ALL the time your oven is on. Peerless-Premier is one brand that is ideal for off-grid homes.

To conserve energy and water when washing clothes, a **front-loading clothes washer** is a must, as is a gas-fired clothes dryer. Better yet, use a clothesline or indoor rack for drying.

Instant water heaters, either gas or electric models, use 20 to 40 percent less energy because they only work when someone turns on the hot water faucet. They also last 30 to 40 years, reducing landfill and resource waste.

Combine a **solar hot water** system with an instant water heater and you've got the lowest-cost, and most ecologically-responsible way to heat domestic hot water.

Energy Conversions

Btu (British Thermal Unit): the energy required to raise one pound of water one degree Fahrenheit.

1 gal. liquid propane = 4 lb. (if you buy propane by the pound)

1 gal. liquid propane = 91,500 Btu

1 gal. liquid propane = 36.3 cubic feet propane gas @ sea level

1 Therm natural gas = 100,000 Btu

1 cubic foot natural gas = 1,000 Btu

1 kW electricity = 3,414.4 Btu/hr

1 horsepower = 2,547 Btu/hr

Source: Solar Living Sourcebook

APPENDIX B
System Sizing

Shopping for a PV/Wind System

When we wrote *Logs, Wind and Sun*, we thought it would be an interesting exercise to request bids from four companies as to what type of system they would recommend given the variables listed on the previous page. Three companies promptly responded to our request; the 4th one took a few weeks. Their responses varied greatly—from asking many questions to making assumptions; from providing very detailed bids to giving an approximate package price.

Their equipment recommendations varied too:
- From 600 watts to 1,440 watts of PV modules (we had 1,140 when surveyed)
- Two inverters (we use one inverter plus a transformer)
- Battery capacity ranged from 1,680-amp hours to 2,340-amp hours (we used the 1,100-amp hour bank for over a year, then added a 2nd one)

Some of the questions they asked: new or existing home, grid-tied or off-the-grid; what elevation; how far from home to solar array; how many gallons of water used per day, wattage of circulating pumps and hours used per day; how many light bulbs and usage per day; number of sunny days per week or month.

Getting the prices for the main equipment is easy, but don't forget to include all the extras: battery cables, fuses and disconnects, wiring, lightning arrestors, shipping, etc. If you are comparing apples to apples, an itemized bid is the only way to go, so there are no surprises in the end. Pick a reputable company: someone who can answer your questions and still be there in a few months when you have more questions.

Steps To Sizing Your System

1. Determine your power consumption (worksheet on page 263).
2. Re-evaluate your consumption to look for ways to conserve. *A rule of thumb that you may hear is that for every dollar you spend replacing inefficient appliances, you'll save $3 in the cost of a renewable energy system.*
3. Find your sun hours per day for your location (page 264 for maps/website).
4. Size your solar array (page 265).
5. Size your battery bank (page 266).

Have this information ready before you request bids. The companies will know you've done your homework and that you are serious about renewable energy.

Our Solar/Wind System
- 1,140-watt PV array, tilted seasonally; plus 480-watt new array (1,620 total)
- 1,000-watt wind turbine, on 50-foot tower
- Trace SW4024 inverter (24-volt)
- Trace 240-volt transformer (to run the well pump)
- Two charge controllers: Easy-Wire Center (wind); OutBack MX60 (solar)
- Two TriMetric meters
- Batteries: primary bank of twenty T-105s (1,100 amp hours); secondary bank of twelve L-16s (1,170 amp hours)
- Trace 250-amp DC Disconnect

Our Home
- Log home is 900 square feet on the main level (with radiant floor heat; temp set at 63 degrees); 600 square feet of open loft (no radiant heat); plus a 1st level basement/garage of 900 square feet (with radiant heat; temp set at 55 degrees)
- Wood stove used extensively for heating the main floor/loft
- Propane used for radiant heat boiler, cooking range, clothes dryer
- Hot water heated with indirect tank connected to main boiler; will add solar hot water soon

Our Power Consumption
The PV/Wind system powers the following:
- Well pump (1.5 hp; draws 11 amps at 240 volts; pulls water from 540 feet at 6 gal/minute; no cistern; 40-gallon pressure tank)
- Radiant heat circulation pumps (5 zones)
- 19 c.f. Kenmore refrigerator (top freezer); plus 5.5 c.f. chest freezer
- Front-loading clothes washer (3 loads/week)
- Dishwasher (2 or 3 cycles per week)
- Microwave, toaster, coffee maker, etc.
- 27" LCD TV (2 to 3 hr/day)
- Computer, 17" flat screen monitor & many peripherals, laser and inkjet printers (6 to 8 hr/day)
- Laptop Computer (6 hr/day)
- Stereo (4 to 5 hr/day)
- All compact fluorescent light bulbs and lots of natural daylight

U.S. Annual Wind Average

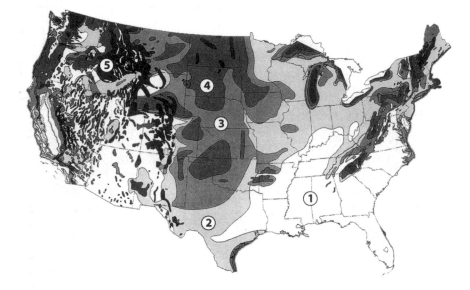

To view this map and individual states in color:
http://rredc.nrel.gov/wind/pubs/atlas/maps.html

Wind Power Class	10 m (33 ft) tower		50 m (164 ft) tower	
	Wind Power W/m²	Speed m/s (mph)	Wind Power W/m²	Speed m/s (mph)
1	0 —	0	0 —	0
2	100 —	4.4 (9.8 mph)	200 —	5.6 (12.5 mph)
3	150 —	5.1 (11.5 mph)	300 —	6.4 (14.3 mph)
4	200 —	5.6 (12.5 mph)	400 —	7.0 (15.7 mph)
5	250 —	6.0 (13.4 mph)	500 —	7.5 (16.8 mph)
6	300 —	6.4 (14.3 mph)	600 —	8.0 (17.9 mph)
7	400 —	7.0 (15.7 mph)	800 —	8.8 (19.7 mph)
	1000 —	9.4 (21.1 mph)	2000 —	11.9 (26.6 mph)

System Sizing — Your Electrical Needs

We recommend doing two calculations: one for winter, and one for summer.

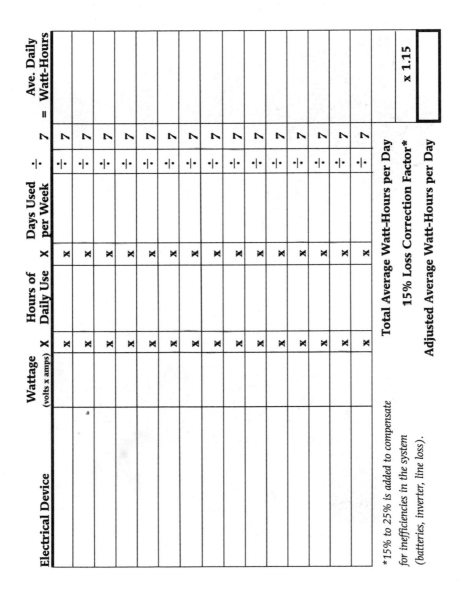

Electrical Device	Wattage (volts x amps)	X	Hours of Daily Use	X	Days Used per Week	÷	7	=	Ave. Daily Watt-Hours
		X		X		÷	7	=	
		X		X		÷	7	=	
		X		X		÷	7	=	
		X		X		÷	7	=	
		X		X		÷	7	=	
		X		X		÷	7	=	
		X		X		÷	7	=	
		X		X		÷	7	=	
		X		X		÷	7	=	
		X		X		÷	7	=	
		X		X		÷	7	=	
		X		X		÷	7	=	
		X		X		÷	7	=	
		X		X		÷	7	=	

Total Average Watt-Hours per Day

15% Loss Correction Factor* x 1.15

Adjusted Average Watt-Hours per Day

*15% to 25% is added to compensate for inefficiencies in the system (batteries, inverter, line loss).

Solar Radiation Maps for Winter & Summer

(Insolation Data)

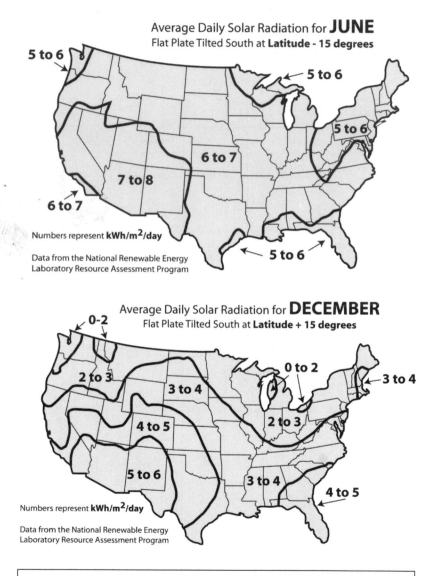

Average Daily Solar Radiation for JUNE
Flat Plate Tilted South at **Latitude - 15 degrees**

5 to 6

5 to 6

5 to 6

6 to 7

7 to 8

6 to 7

5 to 6

Numbers represent **kWh/m²/day**

Data from the National Renewable Energy
Laboratory Resource Assessment Program

Average Daily Solar Radiation for DECEMBER
Flat Plate Tilted South at **Latitude + 15 degrees**

0-2

0 to 2

3 to 4

2 to 3

3 to 4

2 to 3

4 to 5

5 to 6

3 to 4

4 to 5

Numbers represent **kWh/m²/day**

Data from the National Renewable Energy
Laboratory Resource Assessment Program

U.S. Solar Radiation Resource Maps
http://rredc.nrel.gov/solar/old_data/nsrdb/redbook/atlas

Solar Array Sizing Worksheet

		June	Example	December
1.	Input your **Adjusted Ave. Watt-Hours per Day** from page 263.	_____	3000	_____
2.	Find your site on the June and December Insolation Maps on page 264 and input the nearest figure. *See the website, page 264, for more specific information.*	_____	6	_____
3.	To find the number of watts you need to generate per hour of full sun, divide line 1 by line 2.	_____	500	_____
4.	Select a solar module and multiply its rated wattage by .70 (.80 if using an MPPT charge controller).* *Example: Enter 84 for a 120-watt module (or 96 with MPPT).***	_____	84	_____
5.	To find the number of modules needed, divide line 3 by line 4. *Remember that pairs of modules are needed for 24-volt systems; sets of 4 modules for 48-volt systems.*	_____	**6**	_____

This worksheet assumes you will operate entirely on solar power. If you add a wind tower, you can downsize the number of solar modules needed. We suggest to start conservatively, and add more solar modules as needed.

* Another method is to find the rated amperage for a particular module and multiply it by the battery charging voltage (typically 13 volts). Example: Kyocera's 120-watt panel has a Ipmax rating of 7.1 amps x 13 = 92.3 watts. This gives you a slightly more optimistic number.

** You'll get 120 watts from a 120-watt PV module **only** when using an MPPT charge controller during the 2 hours nearest high noon, and **only** when the surface temperature of the module is below 77 degrees Fahrenheit...which is hardly ever. Output is typically derated to 60% - 70% for standard charge controllers (75% - 80% with MPPT) to give you a more accurate number.

Battery Sizing Worksheet

	Average	Example	Winter
1. Input your **Adjusted Ave. Watt-Hours per Day** from page 263.	_____	3000	_____
2. Input the number of days of battery storage you need (the number of cloudy days in a row).*	_____	4	_____
3. Multiply line 2 by line 1.	_____	12,000	_____
4. Input the battery depth of discharge you are comfortable with: 80% discharge (.80) is maximum; 50% discharge (.50) is better.	_____	.50	_____
5. Divide line 3 by line 4.	_____	24,000	_____
6. Derate batteries for low operating temperatures (^{O}F): select a factor next to your lowest operating temperature and enter here: $80^OF - 1.00$ | $70^OF - 1.04$ | $60^OF - 1.11$; $50^OF - 1.19$ | $40^OF - 1.30$	_____	1.11	_____
7. To find your total battery capacity, multiply line 5 by line 6.	_____	26,640	_____
8. Calculate the watt-hour capacity of your selected battery: **voltage x amp hour**. *Example L-16 is 6 volts x 390 amp hours*	_____	2,340	_____
9. To find the number of batteries you need, divide line 7 by line 8. Adjust the number of batteries to fit your system voltage. *Example: 24-volt system requires sets of 4, 6-volt batteries.*	_____	**12**	_____

* If this number is greater than 5, a good backup generator may be more cost effective than extra batteries.

Micro-Hydro Formulas & Conversion Factors

Gross Head x Flow x System Efficiency x C = Power

Example: 50 ft. x .223 cfs x 0.55 x 0.085 = .52 kW (or 520 watts)

- **Gross head** is the actual distance of drop from the intake to the turbine, not taking into account the friction developed in the penstock (in feet or meters).
- **Flow** is measured in cubic feet per second (cfs), or cubic meters per second (cm/s).
- **System efficiency** will be between 40% and 70%.
- **C** (the constant) is 0.085 when using feet; 9.81 when using meters.

Finding the Exact Efficiency of Your System

If you do install an actual system, you can see how close you were to guessing the efficiency factor by running the formula backwards:

$$\text{True Efficiency Factor} = \frac{kW}{\text{Gross Head} \times \text{Flow} \times \text{C (the constant)}}$$

Helpful Conversion Factors

1 cubic foot (cf) = 7.48 gallons
1 cubic foot per second (cfs) = 448.8 gallons per minute (gpm)
1 cubic foot (cf) = 0.028 cubic meters (cm)
1 cubic meter (cm) = 35.3 cubic feet (cf)
1 cubic meter per second (cm/s) = 15,842 gallons per minute (gpm)
1 meter (m) = 3.28 feet
1 foot = 0.3048 meters (m)
1 pound per square inch (psi) = 2.31 feet of head
1 kilowatt (kW) = 1.34 horsepower (hp)
1 horsepower (hp) = 746 watts

Electrical Formulas & Helpful Information

Calculating Line Loss

A while back, I wanted to know what the exact voltage drop would be in the wires if I added more modules to the solar array. By looking at standard wire loss tables I could see that it would be more than 2 percent, and less than 5 percent, but that wasn't close enough. Finally, I found the formula in *Pocket Ref*, a tight little book with more tables and formulas than I ever knew existed.

Following is an abbreviated version. It will work for all DC wiring, and all single phase AC wiring, using copper wire below 121 degrees Fahrenheit. (For 3-phase AC, or aluminum wire, the complete formula can be found in the *Pocket Ref*.)

EXAMPLE:
You want to run 70 feet of #2 wire for a 1,000-watt array, operating at 24 volts. What is the exact voltage drop?

First calculate the amps by dividing the watts (1,000) by the volts (24) = 41.66 amps. Then find the Wire Area in the chart on the next page for #2 wire (66,400). Plug the numbers into the formula:

$$\textbf{Voltage Drop} = \frac{22 \ \times \ 70 \text{ feet} \ \times \ 41.66 \text{ amps}}{66,400}$$

The voltage drop is .966 volts. To find the percentage of loss, divide .966 volts by your system voltage (24 volts) = .040, or 4% loss.
This amperage is possible only if the array is at peak output and you're using an MPPT charge controller. Normal output would be the module's rated amps (Ipmax).

If you used #1 wire instead, the calculations would show a 3.1% loss; 1/0 wire will bring the voltage loss down to 2.5% ... a more acceptable number.

$$\textbf{Actual Voltage Drop} = \frac{22 \ \times \ \text{length of wire in feet} \ \times \ \text{amps}}{\text{wire area in circular mils}}$$

Using the Wire Size/Line Loss Tables

EXAMPLE:
Your 1,100 watt solar array is located 50 feet from the batteries. By referring to the Wire Size/Line Loss tables on the following pages, what size of wire do you need for a 2% loss?

48-volt System	#6 wire
24-volt System	1/0 wire *(over 4 times thicker wire than #6; and proportionately more expensive)*

EXAMPLE:
You have a well pump that draws 10 amps at 24-volts DC, and is 90 feet from the batteries. What size of wire do you need for a 2% loss?

By referring to the 24-Volt chart with 2% voltage drop, 10 amps going 90 feet would require #4 wire. If you settle for a 4% loss, you could use #6 wire, but the efficiency and life of the pump will be decreased.

Pocket Ref states that "Voltage drop should be less than 2% if possible. If the drop is greater than 2%, efficiency of the equipment in the circuit is severely decreased and life of the equipment will be decreased. As an example, if the voltage drop on an incandescent light bulb is 10%, the light output of the bulb decreases over 30%!"

Wire Area in Circular mils	
4/0	212,000
3/0	168,000
2/0	133,000
1/0	106,000
#1	83,700
#2	66,400
#3	52,600
#4	41,700
#5	33,100
#6	26,300

Power Formula

watts = volts x amps

amps = watts / volts

Wire Size / Line Loss Tables
12 Volts : 2% Voltage Drop

12 Volt DC

2% VOLTAGE DROP

Amps in Wire	Wattage at 12V	ONE-WAY DISTANCE FOR VARIOUS WIRE SIZES									
		#14	#12	#10	#8	#6	#4	#2	1/0	2/0	3/0
1	12	45	70	115	180	290	456	720	-	-	-
2	24	22	35	57	90	145	228	360	580	720	912
4	48	10	17	27	45	72	114	180	290	360	456
6	72	7	12	17	30	47	75	120	193	243	305
8	96	5	8	14	22	35	57	90	145	180	228
10	120	4	7	11	18	28	45	72	115	145	183
15	180	3	4	7	12	19	30	48	76	96	122
20	240	■	3	5	9	14	22	36	57	72	91
25	300	■	■	4	7	11	18	29	46	58	73
30	360	■	■	3	6	9	15	24	38	48	61
40	480	■	■	■	4	7	11	18	29	36	45
50	600	■	■	■	■	5	9	14	23	29	36

To Calculate 4% loss, multiply the distance by 2.

■   Exceeds ampacity; do not use wire sizes in this zone; it may cause overheating.

\-   Over 1,000 feet

One-way distances listed: measured from point A (such as the solar array) to point B (the batteries). The **VOLTAGE DROP** refers to the percent of voltage lost due to resistance. All distances calculated in feet for copper wire.

Wire Size / Line Loss Tables
24 Volts : 2% Voltage Drop

24 Volt DC

2% VOLTAGE DROP

Amps in Wire	Wattage at 24V	ONE-WAY DISTANCE FOR VARIOUS WIRE SIZES									
		#14	#12	#10	#8	#6	#4	#2	1/0	2/0	3/0
1	24	90	140	230	360	580	912	-	-	-	-
2	48	45	70	115	180	290	456	720	-	-	-
4	96	20	35	55	90	145	228	360	580	720	912
6	144	15	24	35	60	95	150	240	386	486	610
8	192	11	17	29	45	71	114	180	290	360	456
10	240	9	14	23	36	57	91	145	230	290	366
15	360	6	9	14	24	38	60	96	153	192	244
20	480	■	7	11	18	29	45	72	115	145	183
25	600	■	■	9	14	23	36	58	92	116	146
30	720	■	■	7	12	19	30	48	77	97	122
40	960	■	■	■	9	14	23	36	58	72	91
50	1200	■	■	■	■	11	18	29	46	58	73

To Calculate 4% loss, multiply the distance by 2.

■   Exceeds ampacity; do not use wire sizes in this zone; it may cause overheating.
-   Over 1,000 feet

One-way distances listed: measured from point A (such as the solar array) to point B (the batteries). The **VOLTAGE DROP** refers to the percent of voltage lost due to resistance. All distances calculated in feet for copper wire.

Wire Size / Line Loss Tables
48 Volts : 2% Voltage Drop

48 Volt DC

2% VOLTAGE DROP

Amps in Wire	Wattage at 48V	ONE-WAY DISTANCE FOR VARIOUS WIRE SIZES									
		#14	#12	#10	#8	#6	#4	#2	1/0	2/0	3/0
1	48	180	280	460	720	-	-	-	-	-	-
2	96	90	140	230	360	580	912	-	-	-	-
4	192	40	70	110	180	290	456	720	-	-	-
6	288	30	48	70	120	190	300	480	772	972	-
8	384	22	34	58	90	142	228	360	580	720	912
10	480	18	28	46	72	114	182	290	460	580	732
15	720	12	18	28	48	76	120	192	306	384	488
20	960	■	14	22	36	58	90	144	230	290	366
25	1200	■	■	18	28	46	72	116	184	232	292
30	1440	■	■	14	24	38	60	96	154	194	244
40	1920	■	■	■	18	28	46	72	116	144	182
50	2400	■	■	■	■	22	36	58	92	116	146

To Calculate 4% loss, multiply the distance by 2.

■   **Exceeds ampacity; do not use wire sizes in this zone; it may cause overheating.**
-   **Over 1,000 feet**

One-way distances listed: measured from point A (such as the solar array) to point B (the batteries). The **VOLTAGE DROP** refers to the percent of voltage lost due to resistance. All distances calculated in feet for copper wire.

Resources

The information listed below will give you a good start in the right direction.
The internet is an excellent tool for finding for new information and resources in this
ever-changing business of renewable energy.

RENEWABLE ENERGY

Manufacturers we have mentioned,
or used as reference

African Wind Power
www.africanwindpower.com

Bergey Wind Power
www.bergey.com

Blue Sky Energy, Inc.
www.blueskyenergyinc.com

Bogart Engineering
www.bogartengineering.com

Concorde Battery Corporation
www.concordebattery.com

Controlled Energy Corporation
www.cechot.com

Dankoff (Conergy) Solar Pumps
www.conergy.com

Fronius USA
www.fronius.com

Golden Grain Stoves
www.goldengrainstove.com

Grundfos Pumps Corporation
www.grundfos.com

Harris Hydroelectric
www.harrishydro.com

HS-Tarm
www.woodboilers.com

Jack Rabbit Marine
www.jackrabbitmarine.com

Kyocera Solar
www.kyocerasolar.com

MK Battery
www.mkbattery.com

OutBack Power Systems
www.outbackpower.com

Proven Energy
www.provenenergy.com

PV Powered
www.pvpowered.com

Sharp Solar Systems
http://solar.sharpusa.com

SMA
www.sma-america.com

Southwest Windpower Inc.
www.windenergy.com

Thermo Tehcnologies
www.thermomax.com

Trojan Battery Company
www.trojan-battery.com

Tulikivi
www.tulikivi.com

Vermont Castings Wood Stoves
www.majesticproducts.com

Xantrex Technology
www.xantrex.com

UniRac
www.unirac.com

UNI-SOLAR® Products
www.uni-solar.com

Consultants/Suppliers
contributing to this book

Backwoods Solar Electric Systems
www.backwoodssolar.com

Burnham-Beck & Sun
www.burnhambeck.com

Northwest Energy Storage
www.nwes.com

Real Goods
www.realgoods.com

RMS Electric
www.rmse.com

Sunelco
www.sunelco.com

ORGANIZATIONS

American Solar Energy Society
www.ases.org

Center for Renewable Energy &
Sustainable Technology
www.crest.org

Interstate Renewable Energy Council
www.irecusa.org

The Masonry Heater Association
www.mha-net.org

The Institute for Solar Living
www.solarliving.org

REFERENCE WEBSITES

Database of State Incentives for
Renewable Energy
www.dsireusa.org

Energy Star
EPA energy ratings of appliances
www.energystar.gov

Find A Solar Professional
www.findsolar.com

National Renewable Energy Lab
www.nrel.gov

Sandia's Photovoltaics Program
www.sandia.gov/pv

SolarAccess
www.solaraccess.com

U.S. Department of Energy's EERE
(Energy Efficiency and Renewable Energy)
www.eere.energy.gov

U.S. Solar Radiation Resource Maps
http://rredc.nrel.gov/solar/
old_data/nsrdb/redbook/atlas

Wind Energy Maps/Tables
http://rredc.nrel.gov/wind/pubs/
atlas/maps.html
http://rredc.nrel.gov/wind/pubs/
atlas/tables.html

MAGAZINES

BackHome Magazine
www.backhomemagazine.com

Countryside & Small Stock Journal
www.countrysidemag.com

Home Power
www.homepower.com

Mother Earth News
www.motherearthnews.com

Natural Home & Garden
www.naturalhomemagazine.com

Solar Today
www.solartoday.org

SUGGESTED READING

Chasing the Sun, Nelville Williams

*Natural Capitalism: Creating the
Next Industrial Revolution*
Paul Hawken, Armory Lovins,
L. Hunter Lovins

*The Solar House: Passive Heating
and Cooling,* Dan Chiras

Solar Living Source Book, Real Goods

*There Are No Electrons: Electronics
for Earthlings,* Kenn Amdahl

General Bibliography

Ewing, Rex A. and Doug Pratt. *Got Sun? Go Solar.* Masonville, CO: PixyJack Press, 2005.

Ewing, Rex A. *HYDROGEN—Hot Stuff Cool Science.* Masonville, CO: PixyJack Press, 2004.

Gipe, Paul. *Wind Power for Home and Business.* Post Mills, VT: Chelsea Green Publishing, 1993.

Glover, Thomas J. *Pocket Ref.* Littleton, CO: Sequoia Publishing, 2001.

Gundersen, P, Erik. *The Handy Physics Answer Book.* Detroit, MI: Visible Ink Press, a division of Gale Research, 1999.

Muir, Doris, and Paul Osborne. *The Energy Economics and Thermal Performance of Log Houses.* Gardenvale, Quebec, Canada: Muir Publishing Company Ltd, 1983.

Schaeffer, John and Doug Pratt. *Solar Living Source Book.* Hopland, CA: Gaiam Real Goods, 2001.

Stewart, John W. *How To Make Your Own Solar Electricity.* Blue Ridge Summit, PA: Tab Books Inc., 1979.

Threthewey, Richard, with Don Best. *This Old House Heating, Ventilation and Air Conditioning.* New York: Little, Brown & Company, 1994.

The Solar Electric House. Worthington, MA: New England Solar Electric Inc., 1998.

U.S. Department of Energy. *Home Wind Power.* Charlotte, VT: Garden Way Publishing, 1981.

Glossary

Absorption Stage A stage of the battery-charging process performed by the charge controller, where the batteries are held at the bulk-charging voltage for a specified time period, usually one hour.

Alternating Current (AC) Electric current that reverses its direction of flow at regular intervals, usually many times per second; common household current is AC.

Alternative Energy Energy that is not popularly used and is usually environmentally sound, such as solar or wind energy, hydrogen fuel, or biodiesel. *See also* Renewable Energy.

Amorphous Solar Cell Type of solar cell constructed by using several thin layers of molten silicon. Amorphous solar cells cost less to produce and perform better in sub-optimal lighting conditions, but need more surface area than conventional crystalline cells to produce an equal amount of power.

Ampere (Amp) Unit of electrical current, thus the rate of electron flow. One volt across one ohm of resistance is equal to a current flow of one ampere.

Ampere Hour (AH) A current of one ampere flowing for one hour. Used primarily to rate battery capacity and solar or wind output.

Array *See* Photovoltaic Array.

Battery Electrochemical cells enclosed within a single container and electrically inter-connected in a series / parallel arrangement designed to provide a specific DC operating voltage and current level. Batteries for PV systems are commonly 6- or 12-volts, and are used in 12, 24 or 48-volt operations.

Battery Cell The basic functional unit in a storage battery. It consists of one or more positive electrodes or plates, an electrolyte that permits the passage of charged ions, one or more negative electrodes or plates, and the separators between plates of opposite polarity.

Battery Capacity Total amount of electrical current, expressed in ampere-hours (AH), that a battery can deliver to a load under a specific set of conditions.

Battery Life Period during which a battery is capable of operating at or above its specified capacity or efficiency level. A battery's useful life is generally considered to be over when a fully charged cell can only deliver 80 percent of its rated capacity. Beyond this point, the battery capacity diminishes rapidly. Life may be measured in cycles and/or years, depending on the type of service for which the battery was designed.

Blocking Diode A semiconductor connected in series with a solar module or array, used to prevent the reverse flow of electricity from the battery bank back into the array, when there is little or no solar output. Think of it as a one-way valve that allows electrons to flow forward, but not backward.

Building-integrated PV (BIPV) Where PV is integrated into a building, replacing conventional materials, such as siding, shingles or roofing panels.

Bulk Stage Initial stage of battery charging, where the charge controller allows maximum charging in order to reach the bulk voltage setting.

Cell *See* Photovoltaic Cell.

Cell Efficiency Percentage of electrical energy that a solar cell produces (under optimal conditions) divided by the total amount of solar energy falling on the cell. Typical efficiency for commercial cells is in the range of 12 to 15 percent.

Charge Controller Component located in the circuit between the solar array or wind turbine, and the battery bank. Its job is to bring the batteries to an optimal state of charge, without overcharging them. Most charge controllers have digital displays to help monitor system status and performance. MPPT charge controllers go a step further, by converting excess array voltage into usable amperage.

Circuit A system of conductors connected together for the purpose of carrying an electric current from a generating source, through the devices that use the electricity (the loads), and back to the source.

Circuit Breaker Safety device that shuts off power (i.e. it creates an open circuit) when it senses too much current.

Conductor A material—usually a metal, such as copper—that facilitates the flow of electrons.

Conversion Efficiency *See* Cell Efficiency.

Current Flow of electricity between two points. Measured in amps.

Depth of Discharge (DOD) The ampere-hours removed from a fully charged battery, expressed as a percentage of rated capacity. For example, the removal of 25 ampere-hours from a fully charged 100 ampere-hour rated battery results in a 25-percent depth of discharge. For optimum health in most batteries, the DOD should never exceed 50 percent.

Direct Current (DC) Electrical current that flows in only one direction. It is the type of current produced by solar cells, and the only current that can be stored in a battery.

Distributed System A system installed near where the electricity is used, as opposed to a central system—such as a coal or nuclear power plant—that supplies electricity to the electrical grid. A grid-tied residential solar system is a distributed system.

Electrical Grid A large distribution network—including towers, poles, and transmission lines—that delivers electricity over a wide area.

Electric Circuit *See* Circuit.

Electric Current *See* Current.

Electricity In a practical sense, the controlled flow of electrons through a conductor. In a scientific sense, the non-gravitational and non-nuclear repulsive and attractive forces governing much of the behavior of charged sub-atomic particles.

Electrode A conductor used to lead current into or out of a nonmetallic part of a circuit, such as a battery's positive and negative electrodes.

Electrolyte Fluid used in batteries as the transport medium for positively and negatively charged ions. In lead-acid batteries this is a somewhat diluted sulfuric acid.

Electron Negatively-charged particle. An electrical current is a stream of electrons moving through an electrical conductor.

Energy The capacity for performing work. A ball resting on the top of a hill is said to have potential energy, while the same ball rolling down the hill is imbued with kinetic energy. Solar cells convert electromagnetic energy (light) from the sun into electrical energy, while wind turbines convert the kinetic energy of the air into first mechanical energy, and then electrical energy.

Energy Audit An inspection process that determines how much energy you use in your home, usually accompanied by specific suggestions for saving energy.

Equalization A controlled process of overcharging non-sealed lead-acid batteries, intended to clean lead sulfates from the battery's plates, and restore all cells to an equal state-of-charge.

Evacuated-Tube Collector A type of solar hot-water collector that uses "evacuated," or vacuum-sealed, glass tubes to isolate copper channels through which solar-heated water flows.

Flat-Plate Collector In solar hot-water systems, the primary unit for collecting the sun's energy. Flat-plate collectors consist of a continuous loop of black copper pipe partially embedded in an insulating material within frame, and covered with tempered glass.

Float Stage A battery-charging operation performed by the charge controller in which enough energy is supplied to meet all loads, plus internal component losses, thus always keeping the battery up to full power and ready for service. Float voltage is somewhat lower than bulk voltage.

Fossil Fuels Carbon- and hydrogen-laden fuels formed underground from the remains of long-dead plants and animals. Crude oil, natural gas and coal are fossil fuels.

Full Sun Scientific definition of solar power density received at the surface of the earth at noon on a clear day. Defined as 1,000 watts per square meter (W/m^2). Reality varies from 600 to 1,200 W/m^2, depending on latitude, altitude, and atmospheric purity.

Geothermal Literally, heat from the Earth. This includes the seasonal solar heat stored in the top several feet of the Earth's crust, as well as the heat that naturally conducts toward the surface from the Earth's mantle and core.

Greenhouse Effect A warming effect that occurs when heat from the sun becomes trapped in the Earth's atmosphere due to the heat-absorbing properties of certain (greenhouse) gases.

Greenhouse Gases Gases responsible for trapping heat from the sun within the Earth's atmosphere. Water vapor and carbon dioxide are the most prevalent, but methane, ozone, chlorofluorocarbons and nitrogen oxides are also important greenhouse gases.

Grid *See* Electrical Grid.

Grid-Connected PV System A solar PV system that is tied into the utility's electrical grid. When generating more power than necessary to power all its loads, the system sends the surplus to the grid. At night, the system draws power from the grid.

Ground-Source Heat Pump A type of heat pump that uses the natural heat storage ability of the earth or the groundwater to heat or cool a building.

Heat Exchanger A device used to transfer heat from one reservoir of fluid to another.

Heat Pump *See* Ground-Source Heat Pump.

Hertz (Hz) A unit denoting the frequency of an electromagnetic wave, equal to one cycle per second. In alternating current, the frequency at which the current switches direction. In the U.S. this is usually 60 cycles per second (60 Hz).

Hybrid System Power-generating system consisting of two or more subsystems, such as a wind turbine or diesel generator, and a photovoltaic system.

Hydronic A heating system that uses circulating hot water to transfer heat from a boiler, water heater, or solar collector to the inside of a building.

Insolation Measure of the amount of solar radiation reaching the surface of the Earth. According to NREL, "this term has been generally replaced by solar irradiance because of the confusion of the word with insulation." *See* Irradiance.

Inverter Component that transforms the direct current (DC) flowing from a solar system or battery to alternating current (AC) for use in the home. Also called a power inverter.

Irradiance Rate at which radiant energy arrives at a specific area of the Earth's surface during a specific time interval. Measured in W/m^2.

I-V Curve A graph that plots the current versus the voltage from a solar cell, as the electrical load (or resistance) is increased from short circuit (no load) to open circuit (maximum voltage). The shape of the curve characterizes the cell's performance. Three important points on the I-V curve are the open-circuit voltage, short-circuit current, and peak or maximum power (operating) point.

Junction Box (J-Box) Enclosure on the back of a solar module where it is connected (wired) to other solar modules.

Kilowatt (kW) Unit of electrical power, equal to one thousand watts.

Kilowatt-Hour (kWh) One thousand watts being used over a period of one hour. The kWh is the usual billing unit of energy for utility companies.

Life-Cycle Cost Estimated cost of owning, operating, and disposing of a system over its useful life.

Load Anything that draws power from an electrical circuit.

Maximum Power Point Tracking (MPPT) Technology used by direct grid-tied inverters and some charge controllers to convert, through the use of DC-DC power converters, excess array voltage into usable amperage, by tracking the optimal power point of the I-V curve.

Megawatt (MW) One million watts; 1,000 kilowatts. Commercial power plants and wind farms are usually rated in megawatts.

Micro-Hydro A home-scale hydroelectric system designed to produce electricity by rerouting a portion of a stream's flow through a turbine. The DC power the turbine produces is stored in batteries in the same way solar or wind energy is stored.

Module *See* photovoltaic module.

Monocrystalline Solar Cell Type of solar cell made from a thin slice of a single large silicon crystal. Also known as single-crystal solar cell.

Multicrystalline Solar Cell *See* polycrystalline solar cell.

National Electrical Code (NEC) The U.S. minimum inspection requirements for all types of electrical installations, including solar/wind systems.

NEMA (National Electrical Manufacturers Association) The U.S. trade organization which sets standards for the electrical manufacturing industry.

NREL (National Renewable Energy Laboratory) Based in Golden, Colorado, NREL is the principal research laboratory for the DOE Office of Energy Efficiency and Renewable Energy. Operated by Midwest Research Institute and Battelle, NREL concentrates on studying, testing and developing renewable energy technologies.

Net Metering A practice used in conjunction with a solar- or wind-electric system. The electric utility's meter tracks the home's net power usage, spinning forward when electricity is drawn from the utility, and spinning backward when the solar or wind system is generating more electricity than is currently needed to run the home's loads.

Ohm Measure of the resistance to current flow in electrical circuits, equal to the amount of resistance overcome by one volt in causing one ampere to flow.

Orientation Term used to describe the direction that a solar module or array faces. The two components of orientation are the tilt angle (the number of degrees the panel is raised from the horizontal position) and the aspect angle, (the degree by which the panel deviates from facing due south).

Panel *See* Solar Panel.

Parallel Connection Wiring configuration whereby the current is given more than one path to follow, thus amperage is increased while voltage remains unchanged. In DC systems, parallel wiring is positive to positive (+ to +) and negative to negative (- to -). *See also* Series Connection.

Passive Solar Home Home designed to use sunlight for direct heating and lighting, without circulating pumps or energy conversion systems. This is achieved through the use of energy efficient materials (such as windows, skylights and Trombe walls) and proper design and orientation of the home.

Peak Load Maximum amount of electricity being used at any one point during the day.

Pelton Wheel A type of water wheel used to turn a micro-hydro turbine. A Pelton wheel uses slightly offset cups to catch water and transfer its kinetic energy to the shaft of the turbine.

Penstock In micro-hydro systems, a pipe that conveys water under pressure from the water source to the micro-hydro turbine. Commonly made of PVC pipe.

PEX Tubing Cross-linked polyethylene. Type of tubing used to circulate water or glycol through a floor's hydronic heating system.

Photon Basic unit of light. A photon can act as either a particle or a wave, depending on how it's activity is measured. The shorter the wavelength of a stream of photons, the more energy it possesses. This is why ultraviolet (UV) light is so destructive, while infrared (IR) is not.

Photovoltaic (PV) Refers to the technology of converting sunlight directly into electricity, through the use of photovoltaic (solar) cells.

Photovoltaic Array A system of interconnected PV modules (solar panels) acting together to produce a single electrical output.

Photovoltaic Cell The basic unit of a PV (solar) module. Crystalline photovoltaic cells produce an electrical potential of around 0.5 volts. The higher voltages typical in PV modules are achieved by connecting solar cells together in series.

Photovoltaic Module Collection of solar cells joined as a unit within a single frame. Commonly called a "solar panel."

Photovoltaic System Complete set of interconnected components—including a solar array, inverter, etc.—designed to convert sunlight into usable electricity.

Polycrystalline Solar Cell Type of solar cell made from many small silicon crystals (crystallites). Because of the numerous grain boundaries, devices that employ this design will operate with slightly reduced efficiency. Also known as a multi-crystalline solar cell.

PV Photovoltaic.

Rated Power Nominal power output of an inverter; some units cannot produce rated power continuously.

Renewable Energy (RE) Energy obtained from sources that are essentially inexhaustible (unlike, for example, fossil fuels, of which there is a finite supply). Renewable sources of energy include conventional hydroelectric power, wood, waste, geothermal, wind, photovoltaic, and solar-thermal energy.

Semiconductor Material that has an electrical conductivity in between that of a metal and an insulator. Typical semiconductors for PV cells include silicon, gallium arsenide, copper indium diselenide, and cadmium telluride.

Series Connection A wiring configuration where the current is given but one path to follow, thus increasing voltage without changing the amperage. Series wiring is positive to negative (+ to -) or negative to positive (- to +). *See also* Parallel Connection.

Silicon (Si) The most common semiconductor material used in the manufacture of PV cells. In the periodic table, it is element number 14, positioned between aluminum and phosphorus.

Single-Crystal Silicon *See* Monocrystalline Solar Cell.

Solar Cell *See* Photovoltaic Cell.

Solar Energy Energy from the sun. Virtually all energy on Earth—including solar, wind, hydroelectric and even fossil-fuel energy—originated as solar energy.

Solar Insolation *See* Irradiance.

Solar Module *See* Photovoltaic Module.

Solar Panel Common term used to describe a PV (solar) module. "Solar panel" refers to both photovoltaic modules, used for making electricity, and solar hot-water panels, used to augment a home's heating system. (Compare with flat-plate collector.)

Solar Power *See* Solar Energy.

Stand-Alone A solar-electric system that operates without connection to the utility grid, or another supply of electricity. Typically, unused daylight energy production is stored in a battery bank to provide power at night. Stand-alone systems are used primarily in remote locations, such as mountain areas, ocean platforms or communication towers.

Thin Film *See* Amorphous Solar Cell.

Tilt Angle The angle of inclination of a module measured from the horizontal. The most productive tilt angle is one in which the surface of the module is exactly perpendicular to sun's rays.

Volt (V) A unit of electrical force, analogous to the water pressure within a garden hose. It is equal to the amount of electromotive force that will cause a steady current of one ampere to flow through a resistance of one ohm.

Watt (W) Unit of electrical power used to indicate the rate of energy produced or consumed by an electrical device. One ampere of current flowing at a potential of one volt produces one watt of power. Wind turbines and PV modules are often rated in watts.

Watt-hour (Wh) Unit of energy equal to one watt of power being used or produced for one hour.

Wind Energy The kinetic energy present in wind, measured in watts per square meter (W/m^2). Wind turbines convert the kinetic energy into mechanical energy through the use of propeller blades, which in turn drive an alternator to produce electricity.

Index

italic numbers represent illustrations, graphs, or photos

List of Illustrations

About the Author

Rex Ewing has lived blissfully off-grid with solar and wind energy since 1999 when he left the dusty plains of Colorado and headed for the Rockies to build his wife, LaVonne, a long-promised log home. When he's not writing books or magazine articles about renewable energy—or his first love, horses—he and LaVonne are probably trekking through the back country, canoeing, or enjoying the 50-mile view from their deck.

Before moving to the mountains to concentrate on his writing, Ewing raised grass hay and high-strung Thoroughbred race horses in the Platte River valley. Whenever his employees were clever enough to corral him behind a desk, he served as CEO of a well-respected equine nutrition firm, where he formulated and marketed a successful line of equine supplements worldwide.

Ewing's books include *Got Sun? Go Solar; Power With Nature; HYDROGEN—Hot Stuff Cool Science; Logs, Wind and Sun; and Beyond the Hay Days.* His renewable energy magazine columns can be found in *Log Homes Illustrated* and *Countryside Magazine.*

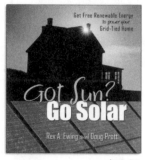